ancient FIRE
modern FIRE

Understanding and Living With
Our Friend & Foe

EINAR JENSEN

PIXYJACK PRESS INC

Published by PixyJack Press, Inc.
PO Box 149, Masonville, CO 80541 USA
www.PixyJackPress.com

print ISBN 978-1-936555-64-2
Kindle ISBN 978-1-936555-65-9
epub ISBN 978-1-936555-66-6

Library of Congress Cataloging-in-Publication Data

Names: Jensen, Einar (Firefighter), author.
Title: Ancient fire, modern fire : understanding and living with
    our friend & foe / Einar Jensen.
Description: Masonville, CO : PixyJack Press, [2015] | Includes
    bibliographical references and index.
Identifiers: LCCN 2015034101 | ISBN 9781936555642
Subjects: LCSH: Fire. | Fire--Social aspects. | Fire--Folklore.
Classification: LCC GN416 .J46 2015 | DDC 363.37/1--dc23
LC record available at http://lccn.loc.gov/2015034101

Front cover photo from iStock.
Interior photos by Kevin Masten, Caroline Jensen, Einar Jensen.
Book design by LaVonne Ewing.

To my family,
who inspires me to explore our world
and share its wondrous past, present and future.

And the fire flaming. And the wheels are turning.

Feel the light and fold in tight.
Feel the wheels turning.

If you have the light, turn off, let's go.
Let's feel the life and forest and fire singing.

In the forest, you feel the fire singing.
You can feel the wheels turning.

You can light the sky entire.
The fire flaming and the wheels are turning.
Fire flaming, down, down, down, down.

Today the night is forming and the wheels
are turning and the fire's flaming!

And the wheels are turning!

Fire night and falling light.

– Tabor and Kali Jensen

Contents

Origins of Fire: Ancient Myths

Prologue

*O*n October 21st, 2007, a boy was at his home along Rocking Horse Road—a ranch near Agua Dulce, a small community on the northeast outskirts of Santa Clarita, California—when he decided to play with matches. Describing his actions as misusing fire is more appropriate, but from his perspective "playing" is probably accurate. He found matches in his home, took them outside and began lighting them, either understanding how to do it already or experimenting long enough to figure out how to rub the match head against the striker panel to create a flame. At some point in his play around one o'clock in the afternoon, he ignited a brush fire that would become known as the Buckweed Fire.

A ridge of high pressure over the Pacific Ocean had created another day of high temperature and low humidity. Warm, dry Santa Ana winds blew between 20 and 40 miles per hour from the ocean across southern California, gusting twice as fast through valleys and down slopes. Buckweed was one of nine large wildfires that ignited in the region that day, starting what would be known as the 2007 California Fire Siege.

Empowered by the gusting wind and fueled by dry chaparral, mixed brush and grass, flames roared out of control, quickly threatening hundreds of homes and causing emergency managers to begin evacuations. Firefighters responded by road and by air, crisscrossing the region as wildfires ignited and expanded, threatening lives and property. The Buckweed Fire burned 10,000 acres by evening and continued churning over the landscape the following days threatening transmission lines, a theme park and thousands of structures. Wind-borne embers sparked spot fires a half mile in front of the fire, leap-frogging containment efforts by firefighters.

Three days after it started, 28 handcrews (each consisting of 20 firefighters), 144 engines (staffed with an average of three firefighters), 13 bulldozers, several helicopters and a management team numbering 130 had contained the wildfire. The boy, through his match and the Buckweed Fire it produced, destroyed 21 homes and 42 other structures and burned over 38,300 acres. Approximately 15,000 people evacuated their communities in advance of the fire.

Less than a month after it ignited, the Los Angeles County district attorney's office announced it would not file charges against the boy because there was no evidence of intent. A *Los Angeles Times* reporter said the boy was "distraught" about the wildfire's consequences. That unsatisfying end to the Buckweed Fire is but one reason why I wrote this manual.

We teach children about the Fire Triangle, but that symbol is only a simple gateway for a complicated topic. Understanding the physical and chemical properties of fire are the topics for Chapter 2. Chapter 3 considers the significance of youth firesetting and what should be done about this growing problem as half of all arson arrests across the country are juveniles. Additional resources for youth misuse of fire are contained in Appendix. Chapter 4 explains the negative consequences of fire, applying the ideas from the previous chapters with concrete examples of destruction. The boy who ignited the Buckweed Fire has a much better understanding of fire's darker side, but he escaped the incident with few consequences.

Destruction is one consequence of fire, and certainly the one that gets the most media time, but its compliment—creation—is equally powerful and lasting. Fire's creative capacity is the topic of Chapter 5. If we humans expect to use such a powerful phenomenon safely, we need to know and follow rules, which is a discussion found in Chapter 6. As the Buddhist quotation that leads this prologue states, fire has sacred meanings that transcend our physical world. Chapter 7 examines spiritual aspects of fire from around the world.

Our unhealthy relationships with fire often spring from how we perceive risk. I address that complicated topic in Chapter 8. If we can understand risk perception better, we can begin making peace with fire. Peace will come from harmony as I explain in Chapter 9. In Chapter 10 I consider consequences of continuing along our current trajectory of burning, injuring and dying. It's an unsustainable trajectory that I hope we change.

Most chapters include excerpts of ancient stories from diverse cultures regarding fire to illustrate historical roots and connections of ideas. The complete stories, which provide a richer context for understanding fire culturally, are in the Origins of Fire: Ancient Myths section, following the chapters.

Paul Gleason, an extraordinary wildland firefighter, implored the rest of us in the fire service to become students of fire. Learning about fire has improved my safety and my understanding of our world, but the lessons aren't simple as Colleen Morton Busch noted in her book, *Fire Monks: Zen Mind Meets Wildfire*:

> *Learning to live with fire is tricky, because there isn't one kind of fire. There are crown fires, slow creeping fires, wind-driven fires, stand-replacement fires, smoldering fires. There are fires in chaparral, fires in pines, fires in oak savannas, fires in buildings made of wood, clay, and stone. There is fire in the center of each human heart. Knowing what kind of fire you live with, a Zen student knows, is an endless, constantly changing, moment-by-moment process.*

If we don't change our understanding of fire, our rules of engagement, or our cultural values, we should expect more tragedies and be willing to pay for them in ever-increasing volumes of dollars, blood, sweat and tears. I'm committed to preventing these tragedies, and I hope to bring more members into my prevention cadre. ❧

1

Fire, Our Friend and Foe

*Then others tried to bring it. The last was the jackrabbit. After
he had stolen the fire, he hid in a thick brush, shek'ei. There he
burrowed. Then he crouched over the fire, holding it in his hands
under his belly. From this the palms of his hands are black. When
he stole the fire it was not extinguished; and so he obtained it for
the people.*

A.L. Kroeber, *Indian Myths of South Central California*, 1907

*Fire is more than an ecological process or an environmental
problem. It is a relationship.*

Stephen Pyne, as quoted in *Fire Monks: Zen Mind Meets Wildfire*

*F*or most people in 21st-century America, fire is easy. Flicking a
lighter, switching on the furnace, and pushing the igniter button on
the grill all produce fire. Wildfires occur out in the woods or on the
nightly news. Movie characters wield fire like a gesture, rarely getting
hurt by it. Restart a video game and evidence of damage and injury
disappears. The dancing flame on a candle's wick mesmerizes with its
simplicity and apparent harmlessness.

Yet fire is not easy. It's never been easy. Using it, living with it,
herding it and extinguishing it are dangerous—often deadly—pur-
suits. Media oversaturation, which seems ubiquitous in our modern
Western culture, makes it difficult for youth and adults to discriminate
between fantasy and reality regarding their perception of fire and how
it behaves. There always has been both a fascination with fire and a
learning curve, but ancestral cultures had less "media" through which

to wade. Ancient tales recall the struggle to acquire it and harness it. Its heat kept the cold at bay while its flickering light kept predators at bay, saving lives. The stories themselves often were told to entertain and educate audiences, providing social and moral guidance regarding this most dangerous and useful of phenomena. Using fire was a rite with great responsibility.

Accidents happened then as they do now, when rules were ignored, forgotten or not known. Explorer Alexander Henry recorded an accidental fire in what is now northeastern North Dakota:

> *Indians came in from the camp below, and even from the upper part of Two Rivers, to inquire into the cause of the conflagration. They supposed that the Sioux had destroyed this fort, and set fire to the grass, as is their custom when they return from war. I was uneasy for some time, fearing the Indians' camp at the hills was destroyed. But the Crees came in with a few skins, and informed us the fire had been lighted at their tents by accident.*

The community of Crees and the Northwest Company representatives with Henry would have dealt with the consequences of that prairie fire, which ignited on December 1, 1800, for many months until spring weather warmed enough for new grass to sprout. Accidents happen, but from my perspective they don't have to happen.

At some point, many humans—especially American ones—transformed fire use into a right with few responsibilities. As a father of two, I strive to teach my daughters about fire's power and how it can both destroy and create. They love using tape to fix things, but fire plays by different rules. Its destruction and creation cannot be undone, even with tape. That lesson is one characteristic of fire too often lost in the modern experience. Our news media suggest that fire's impact is temporary when in fact those impacts become unalterable parts of history.

Other forms of escape such as literature also color our romantic and thus shallow understanding of fire:

> *Red blossom in the winter wood.*
> *red burning in the winter dusk:*
> *windless now, the trees*

stand silent, snow-held, on the turn
and bank of hill where brush,
the limb and twig of axe-slain, fallen tree
burst into final, one last bloom:
orange flower of fire,
blossom of burning bole, of birch
and elm
aflower now this brief last hour
in winter's dusk, in twilight wood,
where hill meets stubbled field
and pasture-land retreats.

Not only is the power of fire lost in these nostalgic lines from August Derleth's "Brush Fire," it also is dwarfed by the power of human-swung axes. Fire isn't a fragile blossom, but it can impact those blossoms. It's also more powerful than a swinging axe.

There were 1.24 million fires reported in the United States in 2013, according to the National Fire Protection Association. Firefighters responded to one fire every 25 seconds. Those fires killed 3,240 civilians, injured over 15,000 people and caused over $11 billion in direct property damage. Humans were responsible for 98 percent of them.

Our general population isn't alone in misunderstanding fire. Firefighters often overestimate predictive models and underestimate fire's power. When the West Fork Fire made a seven-mile run in June 2013 in southern Colorado, the state's media misquoted Eric Morgan, a fire behavior analyst, as describing the extreme fire behavior as "undocumented and unprecedented." As the analyst later qualified, he meant that such a fire had not been recorded in that region. Indeed, such fire behavior may not have been documented, but that doesn't equate to being without precedent. Clearly we need to understand fire better.

After months of nudging along the idea for *Ancient Fire, Modern Fire* in my mind, I decided to base it on a lesson plan I designed for 4th graders. These students are turning 10, which in Colorado makes them legally responsible for fires that they ignite. Other states have set other ages for legal culpability. These children are full consumers of modern media: music, music videos, television shows, online videos, movies, video games, etc. Cognitively and morally they are ripe for learning.

They are hungry to learn social rules and test-drive their potential roles in the world. They need concrete lessons that distinguish between right and wrong rather than abstract lessons that consider the gray areas of an issue. They are starting to conceptualize, learning social rules and developing attitudes about "self." While they don't seem to understand fire as an actual physical phenomenon or cultural force, they are hungry to fill those knowledge gaps.

Having misconceptions about fire, its power and our relationships with it isn't isolated to childhood. As historian Stephen Pyne will illustrate in an upcoming book, 19th-century Americans declared war on wildfire as if it was an enemy that could be vanquished. On August 20, 1886, Captain Moses Harris of the First Cavalry ordered his troopers to put out a forest fire threatening Mammoth Hot Springs at the north end of Yellowstone National Park. Although they did little to influence the fire, their efforts marked the beginning of the federal government's militarized firefighting campaign. Fire prevention posters in the 1940s suggested Japan's leaders would utilize or at least celebrate and benefit from wildfires on American soil. The images also suggested that failing to prevent wildfires made Americans co-conspirators or enemies.

Culturally, as Americans declared war on wildfire, they created a legacy with complex consequences ranging from the development of an extremely expensive military industrial complex to continue the war effort and responsibility-free living for people in enemy-prone ecosystems. In declaring war, our predecessors committed soldiers, equipment and funding in massive amounts to defeat an enemy that preceded our species' presence on the planet.

Unlike organisms, fire cannot be eradicated.

The war on fire has been promoted as a conventional war of heroes battling an enemy. Yet it's become much more of a guerrilla war with all the problems that would be predicted: significant loss of life, destruction of property, psychological injuries, increasing funding with diminishing returns, urges to fight over-aggressively to placate public relations even at the expense of soldiers' lives.

In addition to committing tangible resources to this ill-conceived, multi-generational war, we commit money. Within the U.S. Forest Service alone, the amount of funding budgeted for fighting wildfires has climbed from 16 to 42 percent since 1995, according to a 2014 report.

While firefighting has gained money, other programs have lost money in a process Agriculture Secretary Tom Vilsack called "fire-borrowing." There is significantly less funding available for vegetation and watershed management, research, maintenance, capital improvements and wildlife habitat management. Ironically, many of those projects would indirectly prevent large wildfires and accelerate wildfire recovery.

The War on Wildfire also ravages state and local budgets. In 2014 Colorado's legislators scraped together $20 million from a tight state budget to purchase two fire-spotting airplanes and fund contracts for four helicopters and four single-engine air tankers. Much like the protagonist in *The Butter Battle Book*, with their fingers on the triggers of their triple-sling-jiggers, the legislators now feel much bigger and more capable of vanquishing the enemy. Unfortunately, triple-sling-jiggers are expensive. In a 2014 report written for the governor and state assembly, Paul Cooke, director of the Colorado Division of Fire Prevention and Control, estimated that an average wildfire season in Colorado has 30 large fires that burn 113,000 acres and cost $41.8 million for fire suppression. Colorado's coffers aren't the only ones impacted. In 2013, wildfires occurred in each state and Puerto Rico burning a total of 4.3 million acres. That money has to come from somewhere.

As in other wars, Americans have committed emotional and psychological resources to fighting wildfires. War generates trauma. When wildfires burn homes and kill people—citizens and firefighters alike—fire wins a battle. Yet when fire does win a battle, as with the 2013 Black Forest Fire that burned over 500 Colorado homes and killed two residents, or the Yarnell Hill Fire that destroyed over 100 buildings and killed 19 firefighters, our military firefighting model reacts with escalation, spending more money and drafting more soldiers into the fight after criticizing current strategies and replacing them with new war campaigns.

We have met the enemy and it is us. We have an estranged relationship with fire. At a conference in 2013 Pyne summarized the wildfire challenge specifically as not a fire problem, but a city problem. We create cities and enclaves of urban structures that are vulnerable to wildfire. We develop subdivisions filled with homes made of combustible walls and roofing, ringed with combustible decks. We decorate our properties with enough Austrian pines, junipers, fitzers and piñon pines to emulate western vegetation—some of which was destroyed to make room for the subdivision—and camouflage the neighborhood. We add arborvitae and cedar for more color and drought resistance. In the end we create flammable micro-ecosystems ripe for burning.

Our ill-planned war against fire isn't limited to wildfire. We build our homes from lighter, cheaper pieces of wood treated with carcinogenic chemicals and fill them with synthetic furnishings that burn quicker and hotter than what we used only a few decades ago. We expect immediate response from firefighters but build communities that don't include emergency services in the planning process. We buy firefighters—our ground troops in this battle—bigger vehicles and innovative tools for destroying the beast rather than ones that prevent or limit fire growth. We ignore their long-term injuries: traumatic stress and cancers.

As history has shown, we can't win a war against wildfire or any other fire for that matter. It's time to make peace with it. This book is my effort to start negotiating terms for peace.

Our understanding of fire has changed through time, yet it is a constant pursuit among diverse cultures. As this book explores human relationships with fire, it offers multiple benefits. Life safety educators will find tools for teaching children and adults about structure fires, wildland fires, burns, and youth fire misuse. Teachers will find ancient stories for incorporating into multicultural lessons from the past. General readers also can use it as a guidebook for learning about fire.

Fire hasn't changed from ancient times to today; we humans have changed and adjusted our relationships with fire in the process. Thus far, our modern adjustments have been counterproductive and even dangerous, but I remain hopeful. Despite working in the fire service in both operational and prevention roles since 1998, I'm no expert when it comes to fire. I've looked throughout the world and deep into time for assistance in understanding fire and our cultural relationships with it. *Ancient Fire, Modern Fire* is a broad investigation including stories of both ancient and more recent interactions between humans and fires that resonate with me. I hope this investigation resonates with you. ☇

2

Fundamentals of Fire Science

❦

Biliku [the first person] had a red stone and a pearl shell. She struck them together and obtained fire. She collected firewood and made a fire. She went to sleep. Mite (the bronze-winged dove) came and stole fire. He made a fire for himself. He gave fire to all the people in the village. Afterwards fire was given to all the places. Each village had its own.

Aka-Čari legend

*H*uman cultures throughout the world have their own stories of how they acquired fire and its powers. In most stories, fire simply existed but it was exclusively possessed, guarded and apparently understood by other beings such as deities. That exclusion led to curiosity and envy among humans and their non-human cousins, who typically tried to beg for it, use trickery to steal it and, occasionally, employ brute force to acquire it.

People across the planet share stories of how their ancestors acquired fire but omit details on its physical and chemical properties and rarely included rules for using fire safely. The only rule that mattered—and it is a good one—was that a recipient of fire have integrity and strong character. For example, the Kanien'kehake (or Mohawk) people of what is now northern New York State, received the power to make fire when a young warrior earned it from his guardian animal.

Suddenly a vision came to him, and a gigantic bear stood beside him in the cave. Then Three Arrows heard it say, "Listen well, Mohawk. Your clan spirit has heard your prayer. Tonight

*you will learn a great mystery which will bring help and glad-
ness to all your people."*

*A terrible clash of thunder brought the dazed boy to his
feet as the bear disappeared. He looked from the cave just as a
streak of lightning flashed across the sky in the form of a blazing arrow. Was this the sign from the thunderbird? Suddenly
the air was filled with a fearful sound. A shrill shrieking came
from the ledge just above the cave. It sounded as though mountain lions fought in the storm; yet Three Arrows felt no fear as
he climbed toward the ledge.*

*As his keen eyes grew accustomed to the dim light he saw
that the force of the wind was causing two young balsam trees
to rub violently against each other. The strange noise was
caused by friction, and as he listened and watched fear filled his
heart, for, from where the two trees rubbed together a flash of
lightning showed smoke. Fascinated, he watched until flickers
of flames followed the smoke.*

*He had never seen fire of any kind at close range nor had
any of his people. He scrambled down to the cave and covered
his eyes in dread of this strange magic.*

As a historian, I wonder how much of this tale, or any of the others
I included in this book, has been changed by the person documenting
them. "Strange magic" sounds like a western term, for example, rather
than an endemic phrase. Yet I included these stories, all of which are
included in their complete form after the chapters, to honor the traditions of storytelling and of sharing these insights with new audiences.

Across the continent, along what is now the coast of British Columbia,
the Cowichan people received fire from a generous, gaming bird:

*Our fathers tell us that very long ago our people did not
know the use of fire. They had no need for fire to warm themselves, because they lived in a warm country. They ate their
meats raw or dried by the sun. But after a while their climate
grew colder. They had to build houses for shelter, and they
wished for something to warm their homes...*

"I know your needs," the bird replied, "and I have come to you, bringing the blessings of fire."

"What is fire?" asked all of them.

"Do you see that little flame on my tail?" asked the bird.

"Yes," all answered.

"Well, that is fire. Today each of you must gather a small bunch of pitch wood. With it you can get fire from the flame on my tail. Tomorrow morning I will come here early. Every one of you will meet me here, bringing your pitch wood with you."

Early next morning all arrived at the chosen place, where the bird was awaiting their coming.

"Have you brought your pitch wood?" asked the bird.

"Yes," replied all of the people.

"Well, then," said the bird, "I am ready. But before I go, let me tell you the rules. None of you can obtain my fire unless you obey the rules. You must be persevering, and you must do good deeds. You must strive for the fire, in order that you may think more of it. And none need to expect to get it who has not done some good deed.

"Whoever comes up with me," continued the bird, "and puts his pitch wood on my tail, he will have the fire. Are you all ready?"

Eventually the bird gave fire to a humble woman with a clean ethical record rather than to a man who stole his neighbor's wife. She in turn shared it with her people so they all could warm their homes and cook their food.

On the other hand, there were no rules included in a story from the Yaqui people who lived in the southwestern United States and northwestern Mexico:

Now there is fire in all rocks, in all sticks. But long ago there wasn't any fire in the world, and all of the Yaquis and the animals and the creatures of the sea, everything that lived, gathered in a great council in order to understand why there was no fire.

They knew that somewhere there must be fire, perhaps in

the sea, maybe on some islands, or on the other side of the sea. For this reason, Bobok, the Toad, offered to go get this fire. The Crow offered to help him and also the Roadrunner and the Dog. These four, the winged animals and the dog, went along to help. But Bobok, the Toad, alone, knew how to enter the water of the sea and not die.

The God of Fire would not permit anyone to take his fire away. For this reason he still sends thunderbolts and lightning at anyone who carries light or fire. He is always killing them.

But Bobok entered the house of the God of Fire and stole the fire. He carried it in his mouth, traveling through the waters. Lightning and thunder made a great noise and many flashes. But Bobok came on, safe beneath the waters. Then there formed on the flooding water, little whirlpools of water full of rubbish and driftwood.

Suddenly, not only one toad was seen, but many, many toads swam in the waters. They were all singing and carrying little bits of fire. Bobok had met his sons and had given some fire to one, then another, until every toad had some. These carried fire to the land where they were awaited by the Dog, the Roadrunner, and the Crow. Bobok gave his fire to those who could not enter the water.

The God of Fire saw this and threw lightning at the Crow and the Roadrunner and the Dog. But many toads kept on coming and bearing fire to the world. These animals gave light to all the things in the world. They put it into sticks and rocks. Now men can make fire with a drill because the sticks have fire in them.

The story of Bobok concludes with the suggestion that fire can be had by using sticks and rocks; other stories also suggest that fire lives in sticks and rocks as the following two excerpts from Me-Wuk stories demonstrate:

"All right," answered Wek'-wek, and he sent Koo-loo'-loo to fetch the fire. Koo-loo'-loo shot out swiftly and soon reached the Star-women by the elderberry tree in the far east, in the place

2 – Fundamentals of Fire Science

where the Sun gets up. Here he hid and watched and waited, and when he saw a little spark of fire, he darted in and seized it and brought it back quickly to Wek'-wek and O-let'-te. He held it tight under his chin, and to this day if you look under the Humming-bird's chin you will see the mark of the fire.

Then Wek'-wek asked: "Where shall we put it?"

O-let'-te answered, "Let us put it in oo'-noo, the buckeye tree, where all the people can get it." So they put it in oo'-noo, the buckeye tree, and even now whenever an Indian wants fire he goes to the oo'-noo tree and gets it.

* * *

When Wek-wek the Falcon awoke and saw the fire on Mount Diablo, he knew that Tol-le-loo had stolen the Valley People's fire. So he set out after Tol-le-loo, and eventually caught him. Tol-le-loo denied having taken the fire, and told Wek-wek to search him if he doubted him. Wek-wek searched but could not find the fire because it was inside Tol-le-loo's flute. So Wek-wek tossed Tol-le-loo into some water and let him go on his way.

Tol-le-loo climbed out of the water, and continued east to the mountains, all the while carrying the fire in his flute. Arriving home, he took the fire out of the flute, and placed it on the ground. Then covering it with leaves and pine needles, he wrapped it up in a small bundle.

Le-che-che the Hummingbird and another bird went after it, but they could not catch it and returned empty-handed.

O-la-choo the Coyote-man could smell the fire, and wanted to steal it. He approached the bundle and pushed it with his nose, preparing to swallow it. Suddenly, however, the fire shot up into the sky and became the Sun.

The people took the fire that was left and put it into two trees, the buckeye and the incense cedar, where legend says it still resides. From that time on, the Mountain People made their fire drills from the wood of these two trees.

Because the laws of physics are universal, the need to rub special sticks together in a specific technique wasn't limited to North America. Maui, the trickster figure of Polynesian tales, learned how to make fire after a game of wits with a mud hen:

> *Maui worked hard, but not a spark of fire appeared. Again he caught his prisoner by the head and wrung her neck, and she named a kind of dry wood. Maui rubbed the sticks together, but they only became warm. The neck-twisting process was re-sumed and repeated again and again, until Maui had tried every tree and the mud hen was almost dead. At last Maui found fire. Then as the flames rose he said, "There is one more thing to rub." He took a fire stick and rubbed the top of the head of his prisoner until the feathers fell off and the raw flesh appeared. Thus the Hawaiian mud-hen and her descendants have ever since had bald heads, and the Hawaiians have had the secret of fire making.*

As Maui learned through trial and error, you can't just stroke a couple sticks together or beat some rocks against each other to make fire. There is technique involved because of fire's physical and chemical properties.

The Chemistry of Fire

Although fire is essential to human culture, it occurs on our planet independent of humans. Lightning, volcanic magma, and meteoric chunks of rock that find their way through the atmosphere have caused fires since the planet first formed because they encountered fuels and an oxygen-rich atmosphere. Those three ingredients—heat, fuel and oxygen—are needed for fire, or more accurately combustion, to occur. We

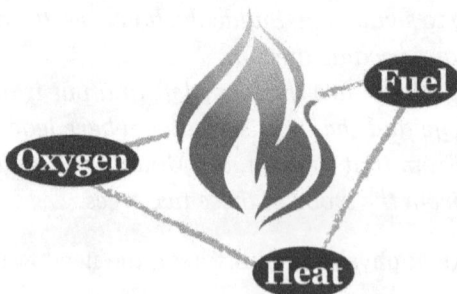

use a triangle, the fire triangle, to symbolize the relationship between heat, fuel and oxygen.

The air on our planet consists of approximately 21 percent oxygen, which exceeds the 16 percent level required by combustion in which oxygen reacts with a substance to release heat and light. Heat needed to create a fire comes from a variety of sources such as magma, sunlight focused by a magnifying glass, a spark of static electricity, and friction.

Fuel is any combustible material, but before any fuel can ignite there must be enough heat to cause the remaining water vapor at the site of heat exposure to evaporate and even more heat to transform the liquid or solid fuel into its vaporous state, a process called pyrolysis. There also must be enough vapor present; too much or too little, which are amounts beyond the upper and lower flammable limits for specific atmospheric conditions, won't burn. When the right amount of this vapor ignites in the presence of the fuel's minimum ignition energy (usually a fraction of a millijoule) and burns, it produces light-producing flames. Scientists refer to the evolving chemical reaction that produces variable intensities of light and heat as rapid oxidation while the rest of us refer to the fuel as being "on fire."

The chemical reaction of this rapid oxidation process is $CH_4 + 2O_2 = CO_2 + 2H_2O + Heat$. Oxygen is the limiting reagent. Without enough oxygen, burning is incomplete and produces more byproducts than complete burning. The actual reaction involves at least 31 steps. Light and heat are always produced, but a glitch in any step produces other byproducts such as smoke and toxic gases.

Light is one of fire's most attractive characteristics from both sacred and profane perspectives. In our electrified society, a single flame is interesting to watch, but doesn't produce much light. When I guided tours in an underground silver mine several years ago, I'd shut off the lights deep in the tunnel, light a single match and ignite a candle to demonstrate how much light 19th-century hard rock miners used when working underground. Compared to the surrounding darkness, a single flickering flame was intense and comforting.

That light as well as individual colors are produced through chemical reactions on molecular and atomic levels. A candle, for example, consists of carbon and hydrogen atoms locked together to form its wax and the wick. Oxygen molecules in the surrounding air bounce against the wax and wick with little consequence. When a heat source is applied to the wick, hydrogen and carbon bonds break and the wax vaporizes (pyrolysis). Many of these liberated atoms rearrange and create new bonds with the surrounding oxygen to form dihydrogen monoxide (also known as H_2O or water) and carbon dioxide.

Both reactions release blue light. Single carbon atoms remain in the air and consolidate into larger particles called soot. Soot absorbs heat energy that in turn converts into light, which causes soot to glow yellow, orange and red. The glowing soot particles and chemical reactions are called flames. Efficient burning produces blue light; inefficient burning produces yellow, orange and red.

A Source of Heat

Flames are more than light. They are hot. Understanding heat energy requires a short summary of thermodynamics. Temperature is a measurement of kinetic energy in a sample of gas, liquid or solid. Heat, on the other hand, is the amount of energy transferred from one object to another because of a difference in temperature. In fact, the most basic principle of thermodynamics is that if two objects are the same temperature, a condition known as thermal equilibrium, heat will not flow between them. Otherwise heat flows from hotter objects to colder objects.

Additionally, energy is conserved in the universe. It cannot be created or destroyed, but it can be used and changed. When any energy is used, some of it becomes heat. Once energy becomes heat, it is most difficult to convert into a more useful form of energy. As a result, the amount of useful energy slowly shrinks, a phenomenon that physicist Rudolf Clausius termed entropy.

Because fire releases heat, it is called an exothermic reaction. The amount of heat that a fuel can produce, measured as British Thermal Units (Btu) per kilogram or pound, is called its heat of combustion. Different fuels release different amounts of energy, which influences a number of human activities, as well as fire growth and suppression.

When heat energy moves, it does so in one of three ways: convection, radiation or conduction. Understanding each form of heat movement improves our understanding of how fires burn as well as how to prevent fires from burning us and our stuff.

Imagine a cup of hot cocoa. If you place your palm a few inches above the cup, your hand feels warmth because a column of hot air and water vapor (steam) is rising above the hot liquid. When a gas or liquid moves in response to a temperature difference—in this case, warmer and less dense air moving upward—it is called convection.

Now place both hands a couple inches away from the sides of the cup. Again, your hands feel warmth, but this form of heat transfer is called radiation. When waves of radiant heat (more accurately infrared electromagnetic radiation) travel from a liquid or solid through the air, they don't heat the air itself. Instead the waves are absorbed by another liquid or solid object in line of sight with the heat source, and the energy is converted to heat.

Finally, put a metal spoon in the cocoa for a couple minutes and the spoon warms. Heat from the liquid has moved into and through the spoon from the warmer end to the cooler end in a movement called conduction. This heat energy transfers from molecule to molecule, based on the density of the material. Conduction is most prevalent in solids, but can occur in liquids and gases.

Convection, radiation and conduction—thermodynamics—explain why smoke detectors and sprinkler heads are placed on ceilings: because hot air rises. They explain why junipers and fitzers are bad plants to have within 30 feet of homes: because enough heat can radiate from those sap-filled "little green gas cans" to ignite wood siding. They explain why using wet gloves to handle hot metal is a bad idea: because heat from the metal conducts quickly through soggy leather to skin, causing significant pain. They also explain how a large wildfire can generate its own weather.

Convection explains cloud formation generally. Heating along the Earth's surface destabilizes the atmosphere because as parcels of air warm up, they rise until they encounter air of the same or higher temperature. The greater the temperature difference, the faster the air moves upward. Signs of atmospheric instability include dust devils, which look like small tornadoes that rise from the surface, and fire whirls, which look like burning cyclones.

When rising air columns contain water vapor, cumulonimbus clouds form. As rising air cools, it loses its capacity to hold water vapor. That abandoned water vapor condenses and creates clouds. Cloud formation continues as long as warmer, saturated air continues rising, off-loading water vapor and releasing heat energy. Condensation is a reverse of the process of vaporization. Every pound of water vapor that condenses back to liquid releases the almost 1,000 Btu of energy that was required to vaporize it in the first place. As more water vapor condenses in the atmosphere, an enormous amount of energy is released. This energy is the primary driver of weather phenomena such as thunderstorms and even hurricanes. When the weight of the water droplets is heavier than the convective column or updraft, rain falls and drags air with it to create downdraft winds. This process is apropos here because it can occur above large wildfires, responding to and influencing their growth.

Smoke, which consists of gases such as carbon dioxide and carbon monoxide, water vapor, and gazillions of tiny partially burned particles of fuel, rises above a fire when it is hotter than the surrounding air. As rising smoke cools, it also loses its capacity to hold water vapor and fuel vapor. Clouds form. If the weight of the water vapor and fuel vapor overwhelms the convective column, or grows so dense that winds aloft in the atmosphere can push it, the smoke column can collapse vertically or horizontally. A collapsing plume dumps superheated air and vaporized fuel onto an already volatile surface to create infernos.

Depending on the amount of air in the convection column, a fire may have the potential to lift and throw embers—chunks of burning material of varying sizes—hundreds and sometimes several thousand feet downwind of the main fire to create spot fires. Wildfires can produce a blizzard of sand- and pea-sized embers immediately downwind, but larger wildfires can throw burning pinecones, shingles and even softball-sized embers beyond the perimeter of the main fire, complicating safety and suppression efforts for firefighters and civilians alike.

Investigators in Texas used damage to trampolines to illustrate the volume of embers generated by the Bastrop County Complex Fire in 2011. While ember production varied based on fuel characteristics, fire behavior and weather, their trampoline research suggests that a blizzard is the appropriate adjective to describe how much burning debris can be in the air. Seven trampolines caught between 82 and over 6,000 embers. The trampoline with the least damage was at a property where the house survived; homes associated with the other six did not.

Effects of Smoke

As established earlier, fire is a series of oxidations: chemical combinations of elements with oxygen. When carbon-based fuels burn, they release carbon atoms that quickly combine with oxygen. When one carbon atom joins two oxygen atoms, carbon dioxide (CO_2) results; it sounds benign, but CO_2 can be dangerous. Clearly our bodies don't need it if we exhale it, but how toxic could it be?

Carbon dioxide is significantly toxic in high doses or extended exposures. It can occur even when the air contains enough oxygen to sustain life. Toxic levels of CO_2 cause high blood pressure, flushed skin, headache, and muscle twitching. Irregular heartbeats, vomiting and hallucinations are possible with higher levels, as is death. It's dangerous because it displaces the amount of oxygen in the body, a condition called asphyxiation or anoxia. The CO_2 replaces O_2 in the bloodstream and in tissues.

Sometimes only one oxygen atom is available for a liberated carbon atom and that reaction produces carbon monoxide (CO). Flaming fires in oxygen-rich environments produce low amounts of carbon monoxide because there is plenty of oxygen. Oxygen-poor environments, however, such as those supporting smoldering fires, produce higher amounts of CO.

Carbon monoxide is a deadly byproduct of fire. It is odorless and colorless and difficult to notice in small quantities without mechanical detectors. Firefighters refer to CO as a narcotic or asphyxiant gas because it depresses the central nervous system, which results in reduced awareness, intoxication, loss of consciousness and, in the worst cases, death.

Hemoglobin, a protein in red blood cells, carries oxygen from our lungs to our cells in a loose chemical combination called oxyhemoglobin. Carbon monoxide interrupts that process, adhering to hemoglobin roughly 200 times more readily than oxygen, forming carboxyhemoglobin that remains in the bloodstream for several years. As cells receive carbon monoxide instead of oxygen, they starve and die.

Although an individual's general physical condition—age, physical activity and duration of exposure—influence the actual carboxyhemoglobin level in the individual's blood, concentrations of the gas above five-hundredths of one percent (500 parts per million) are considered dangerous. When the level increases above one percent, unconsciousness and death can occur without warning.

Flames get the media attention, yet smoke is often the deadliest byproduct of fire. As described earlier, it consists of tiny partially burned particles of fuel, varying gases and water vapor. Aside from the water, most of its contents are toxic and combustible. According to researchers at the University Corporation for Atmospheric Research (UCAR), wildfire smoke often contains aldehydes, nitrogen oxides, ozone, polynuclear aromatic hydrocarbons, volatile organic compounds, heavy metals, and carbon monoxide. Smoke from structure fires also contains substances such as hydrogen cyanide, acrolein, benzene, and hydrogen chloride. As a result, smoke can irritate our eyes, mucous membranes, and lungs. It can also contribute to cancer, injure skin tissue, and kill us.

"Even with his goggles on," Colleen Morton Busch wrote of a man battling a wildfire in California, "his eyes watered."

> *The smoke found the smallest opening. His bandanna filtered some of it, but not enough. He could taste smoke, smell it, feel it filling his nostrils, coating his throat. He felt dizzy, sick to his stomach. Then the coughing started. At first he covered his mouth with his elbow, a relic of zendo decorum... but soon the coughing took over. It wasn't the kind of cough you could cover. It was sputtering, full-body spasm.*

He was coughing from the dirty air, to be sure, but also because the smoke was burning his throat and lungs.

Smoke is wicked hot, as I tell my students. Heat released during combustion yields additional molecules of fuel that in turn yield additional heat. Those molecules also absorb heat energy and radiate it to relatively cooler surfaces as they flow away from the fire. Within a building, the smoke column rises until it encounters a ceiling and radiates outward along paths of least resistance. After covering the entire ceiling, smoke begins filling the room from the top to the bottom similar to the way water fills a bathtub (but upside down). The growing layer of smoke radiates heat via infrared electromagnetic radiation toward other fuels, which can absorb enough heat to lose their water vapor, reach their points of ignition, and vaporize, thus adding more heated combustible fuel to an incendiary environment.

Flashover, Flameover and Backdraft

When this heating and vaporization occur within an enclosed space, enough fuel vapor may be produced to overwhelm the supply of oxygen. Known as a ventilation-controlled fire, this condition can lead to flashover. Flashover is a transitional phase in a fire within an enclosed compartment in which surfaces exposed to thermal radiation reach their ignition temperatures nearly simultaneously allowing fire to spread rapidly throughout the space. Flashover completely ignites an enclosed space. Nothing survives this phase of development.

Flashover is often confused with flameover, which is also called rollover. Unlike flashover, it's not deadly. Firefighters most often observe flameover because it consists of flames rolling through superheated particles of smoke in billowing waves.

Ventilation-controlled fires also can produce backdrafts. Limited oxygen can produce large volumes of unburned vaporized fuel. If oxygen is introduced suddenly, it can mix rapidly with the fuel, creating a highly combustible mixture of gases. If a heat source with the minimum ignition temperature comes in contact with that combustible mixture, extremely rapid burning can result, forcing vaporized fuel out the opening through which oxygen had just entered and creating a fireball or explosion in the oxygen-rich air outside the enclosure. Backdrafts are dangerous because they explode outward beyond the burning room, often surprising occupants and firefighters.

Smoke fresh from a structure, wildland or vehicle fire is hot enough to burn the inside of your mouth, nose and throat on its way to your lungs. When your throat burns, its tissue swells. With enough swelling, a throat can close entirely, suffocating the victim. That's why firefighters implore people to crawl low and go.

Crawling under the hottest layers of air in relatively clean and cool air is the best escape route. We firefighters also crawl low because the layers of heat can be hot enough to melt our helmets and ignite our fire-resistant clothing. Wicked hot.

Firefighters Tricks of the Trade

Firefighters may seem like magicians at times, but they use physics to extinguish fire. Traditional reactive firefighters, the men and women who respond to fires after ignition, are best known for spraying water onto flames. Liquid water absorbs heat from the fire and evaporates, forming water vapor.

The conversion of liquid water to vapor (steam) absorbs large amounts of energy. A single Btu (British Thermal Unit) is the amount of heat energy required to raise the temperature of one pound of water one degree Fahrenheit. However, converting that same pound of water to vapor requires almost 1,000 Btus. If the supply of liquid water is greater than the supply of heat, it can absorb the fire's heat and thus extinguish the fire.

An essential component of heat absorption is surface area. The greater the surface area of a given volume of water, the faster it can absorb heat. Firefighters use fog streams and broken streams of water to create zillions of water droplets that have vastly greater surface areas to absorb heat. As those droplets absorb energy, they convert into steam. Sprinkler systems are effective for similar reasons. After water discharges from the piping onto a deflector, the water stream breaks into droplets capable of absorbing more heat and containing a fire's growth.

Firefighters also attack the other legs of the triangle. They smother fires by preventing oxygen from reaching the fire. They spray foam onto petroleum-based fires to block oxygen from reaching the hot oil, gasoline or grease. That's the premise of commercial hood systems usually found in restaurants. When grease-laden vapors ignite on surfaces protected by these systems, foam sprays from nozzles to create a bubbly blanket on the grill or fryer. As long as the blanket of foam remains intact—blocking oxygen from reaching the hot vapor—the fire dies. If the foam is breached and enough heat remains, flames can reignite. Stop, Drop and Roll also is based on breaking the oxygen leg; rolling on the ground blocks oxygen's access to the hot fuel so flames cease.

Additionally, firefighters remove fuels from fires. Urban firefighters once used pike poles and explosives to demolish buildings in front of advancing conflagrations. Wildland firefighters use specialized hand-tools and fire itself to create vegetation-free strips in front of wildfires. In each case, these firefighters are removing at least one leg—heat, fuel or oxygen—of the fire triangle, which causes the fire to stop burning.

Proactive firefighters use physics to prevent fires from igniting and limit damage from fires that do begin. They enforce codes that require construction materials to be ignition resistant and decorations to produce less heat. They engineer rooms so that walls, ceilings, floors and doors restrict air flow to prevent fires from having an unchecked supply of oxygen. Our current teaching includes reminders to close doors to prevent wind-driven fires from burning through homes any more quickly than normal. They encourage homeowners to use fire-resistant plants when landscaping around homes. Modern firefighting combines both strategies, proactive and reactive, to prevent a fire triangle from forming and break any triangles that do form.

Firefighting Garb

Firefighters utilize physics to protect themselves from fire. The men and women who choose to engage fire in order to protect lives and property must wear clothing capable of protecting them from all its hazardous facets.

Initially firefighters wore their regular clothes and stayed far enough away from the heat to avoid death. They weren't able to enter smoke-filled environments—or at least stay within them for long—until they found technology to protect their airways. A full history of the personal protective equipment (PPE) utilized by firefighters isn't necessary here, but a summary of their current bunker or turnout gear is useful.

Firefighters strive to protect their entire bodies from fire because any exposed skin is a potential vulnerability. As we tell children during presentations, our PPE is fire-resistant but not fire-proof. It's also an elaborate package of technology.

The pants and coat consist of three barriers, protecting firefighters from flames, vapors and heat. The outer layer is flame-resistant while the innermost layer is a vapor barrier. It blocks flammable vapors from reaching our skin, yet it also traps the vapor from our own sweat against us and causes steam burns. Finally a layer of air between the outer shell and vapor barrier acts to insulate firefighters from a fire's heat.

A firefighter's helmet protects the head from heat, objects falling from above and hot water. The helmet's elongated back channels water onto the person's back rather than under their coat where it could convert to steam and cause burns.

Firefighters protect their hands from heat and sharp objects with gloves. They use goggles and face shields to protect their eyes. They utilize a fire-resistant hood to protect their ears, scalp, hair and neck from heat. Boots protect feet from heat, injuries from falling objects, and sharp objects that might poke through the floor.

They protect their airways with a shatter-proof mask connected to a bottle of air. The self-contained breathing apparatus completes an ensemble that can look quite intimidating, especially in a scary environment such as a burning home. SCBA contains compressed air within a bottle, often made of carbon fibers. The bottle rests on a backpack or harness and is connected to the mask through an aid hose and a regulator. The SCBA also holds a device that creates a high-pitched alarm when the device stops moving. This PASS device alerts other firefighters when a crew member stops moving because of death, injury or entrapment.

Thus we teach children (and adults) that firefighters are regular people who have to wear special clothing to stay safe. Firefighters are friends even when they are covered by the fancy equipment.

Firefighters wear other PPE for specialized fire environments. They wear lightweight fire-resistant clothing for wildfires so they can remain mobile and work longer shifts. They wear suits that resemble aluminum foil when responding to aircraft fires or oil-well fires. Those suits are capable of reflecting the extreme heat of burning jet fuel or raw petroleum.

Heat Sources

Heat sources that provide minimum ignition energy are prevalent. They can be autogenous or piloted. Autogenous heat is produced by vibrating molecules during a material's chemical decomposition (pyrolysis), while piloted heat sources are introduced into the situation.

Friction is an underrated heat source that can ignite a fire. While the image is tied to rubbing two sticks together, it's also the heat source responsible for the function of matches. Simply, friction is a force that acts against the sliding movement of two solid objects. It always produces heat.

Because all surfaces have irregularities, including some that are beyond the abilities of our eyes to perceive because they occur on a molecular level, they don't slide against each other perfectly. As the low spots and high spots of those surfaces rub against each other, they attract molecules from the other surface and form new chemical bonds. As the surfaces continue sliding, those bonds break. Creating and breaking chemical bonds uses energy and creates heat. Energy can't be created or destroyed, as established earlier, so friction transforms kinetic energy, which is useful, into thermal energy (heat), which isn't necessarily as useful unless you're trying to create a fire.

Invented in 1827, matches have evolved into two types: safety and strike-anywhere. A safety match earned its name from needing a specific type of surface against which to be rubbed. That surface, which is placed on the match book or box away from the match heads, consists of sand, powdered glass and red phosphorus. The match head consists of sulfur, glass powder and an oxidizing agent—typically potassium chlorate—that releases oxygen when it burns. When the match head is dragged against the striking surface with sufficient force, friction between the powdered-glass surfaces generates enough heat to convert red phosphorus into white phosphorous. As an exothermic reaction that releases heat, that transformation causes the potassium chlorate to decompose and produce oxygen. Oxygen atoms excitedly bounce against the sulfur and, because heat is present, create a flame. Heat from this burning sulfur quickly dries and vaporizes the matchstick so that it can ignite and sustain a flame until the fire runs out of fuel.

A "strike anywhere" match differs only in that the red phosphorus is added to the match head. Thus it can be struck on any surface capable of producing heat from friction and initializing the needed chemical reactions.

Friction also gets credit for flint and steel ignitions. Friction between the sharp edge of a piece of flint and a steel rod rapidly heats and breaks off a piece of steel. We refer to that glowing hot steel shard as a spark. When it lands on a fine fuel—any particle with a large surface area-to-volume ratio, such as dryer lint or vegetable fiber (dry grass)—the spark can transfer enough heat to ignite a fire.

Because it's the steel that creates the spark, other hard stones can be used as long as the striker is crafted into a blade. Agate, quartz and even jade are hard enough to generate a hot spark from the metal.

Thus kindling and tending a fire isn't magic; it's applied science. U.S. Army Captain Randolph Marcy offered instructions on setting fires based on his own knowledge of fire science in *The Prairie Traveler: A Hand-book for Overland Expeditions*, a guidebook for westbound emigrants he wrote in 1859:

> *It is highly important that travelers should know the different methods that may be resorted to for kindling fires upon a march.*
>
> *I have seen an Indian start a fire with flint and steel after others had failed to do it with matches. This was during a heavy rain, when almost all available fuel had become wet. On such occasions dry fuel may generally be obtained under logs, rocks, or leaning trees.*
>
> *The inner bark of some dry trees, cedar for instance, is excellent to kindle a fire. The bark is rubbed in the hand until the fibres are made fine and loose, when it takes fire easily; dry grass or leaves are also good. After a sufficient quantity of small kindling fuel has been collected, a moistened rag is rubbed with powder, and a spark struck into it with a flint and steel, which will ignite it; this is then placed in the centre of the loose nest of inflammable material, and whirled around in the air until it bursts out into a flame. When it is raining, the blaze should be laid upon the dryest spot that can be found, a blanket held over it to keep off the water, and it is fed with very small bits of dry wood and shavings until it has gained sufficient strength to burn the larger damp wood. When no dry place can be found, the fire may be started in a kettle or frying-pan, and afterward transferred to the ground.*
>
> *Should there be no other means of starting a fire, it can always be made with a gun or pistol, by placing upon the ground a rag saturated with damp powder, and a little dry powder sprinkled over it. The gun or pistol is then (uncharged) placed with the cone directly over and near the rag, and a cap explod-*

ed, which will invariably ignite it. Another method is by plac-ing about one fourth of a charge of powder into a gun, pushing a rag down loosely upon it, and firing it out with the muzzle down near the ground, which ignites the rag.

The most difficult of all methods of making a fire, but one that is practiced by some of the Western Indians, is by friction between two pieces of wood. I had often heard of this process, but never gave credit to its practicability until I saw the exper-iment successfully tried. It was done in the following manner: Two dried stalks of the Mexican soap plant, about three fourths of an inch in diameter, were selected, and one of them made flat on one side; near the edge of this flat surface a very small indentation was made to receive the end of the other stick, and a groove cut from this down the side. The other stick is cut with a rounded end, and placed upright upon the first. One man then holds the horizontal piece upon the ground, while another takes the vertical stick between the palms of his hands, and turns it back and forth as rapidly as possible, at the same time press-ing forcibly down upon it. The point of the upright stick wears away the indentation into a fine powder, which runs off to the ground in the groove that has been cut; after a time it begins to smoke, and by continued friction it will at length take fire.

This is an operation that is difficult, and requires practice; but if a drill-stick is used with a cord placed around the centre of the upright stick, it can be turned much more rapidly than with the hands, and the fire produced more readily. The up-right stick may be of any hard, dry wood, but the lower hori-zontal stick must be of a soft, inflammable nature, such as pine, cottonwood, or black walnut, and it must be perfectly dry. The Indians work the sticks with the palms of the hands, holding the lower piece between the feet; but it is better to have a man to hold the lower piece while another man works the drill-bow.

It's not magic at all, but as science it requires us to learn actively. That learning is best and safest when experts teach novices. Leaving novices to learn on their own can produce tragic consequences. ☙

3

Youth and Firesetting:
Playing with Fire Can Burn Us

All burning was limited to certain Indians who were looked up to as leaders or who understood how it should be handled.

Me-Wuk Chief, as recorded by the USFS in 1935

If fire breaks out, and catch in thorns, so that the stack of corn, or the standing corn, or the field, be consumed therewith; he that kindled the fire shall surely make restitution.

Exodus 22:6, King James Version of *The Bible*

*E*arly in my professional community risk reduction career, I earned my certification as a Juvenile Firesetter Intervention Specialist. The whole topic of children willfully misusing fire was taboo in the first agency I worked for. Kids in that wholesome mountain community didn't have such character flaws or dysfunctional, anti-social behavior. That was the company line I heard even as our firefighters responded to fires ignited in a bathroom of the local high school.

Contrary to this not-in-my-backyard bias, children do misuse fire regardless of where they live. They do so for simple reasons such as curiosity and complex reasons such as reacting to psychological stressors, which is why communities should have personnel trained as juvenile firesetter intervention specialists.

According to ancient stories, acquiring fire was reserved for individuals of character capable of creative problem solving. Instead of finding a box of matches in a drawer or old fireworks on a garage shelf, individuals had to travel great distances and confront powerful beings to find

36

or steal fire. In this Anishinaabe story, Naanabozho—the heroic trickster figure—asked his grandmother how to get ishkode (Anishinaabe for fire) to save them from facing another chilly night in the upper Midwest:

> ..."Nookomis, it is very cold here at night. I wish we had ishkode to warm ourselves."
>
> Nookomis said, "Nia, my grandchild, I don't think we can get ishkode as it is a long distance across anishinaabe gichigami to the island home of giizis manidoo, sun spirit, where ishkode can be found, and then too, a person would have to go before one of his smiles, but that has never been done for the island and the sacred lodge of giizis manidoo is strongly guarded by many manidoog and his only companions are his daughters—clouds—who weep rain for him when he is angry or when his heart feels bad.
>
> "No, my grandchild, I would not advise you to undertake such a long journey and so hazardous a task."

Acquiring fire today is much easier; in many cases, it's too easy.

Nationally, children misusing fires playfully or malignantly cause approximately 8,000 residential fires, 700 injuries, 100 deaths and $200 million in property loss annually, according to the National Fire Protection Association's data analysis. Additionally, NFPA researchers found that:

- Lighters and matches accounted for 65 percent of child-misuse home structure fires, 80 percent of associated civilian deaths and 81 percent of associated civilian injuries;
- Children aged five and younger are responsible for 47 percent of home structure fires that are ignited by misusing fire;
- 65 percent of all fatalities of child-misuse fires were set by children aged five and younger;
- Youth fire-misuse caused approximately 10,000 outside fires each year from 2003 to 2006, resulting in annual averages of one civilian death, 96 civilian injuries and $3.3 million in direct property damage;

- Youth between the ages of 11 and 14 are at the greatest risk for setting fires;
- Nearly half of all arson arrests in the United States, according to the FBI, are children under the age of 18.

These statistics suggest that some children need more supervision and less access to lighters and matches, some parents need more information on how to teach safe fire skills, and both children and adults use fire deliberately rather than accidentally.

They also suggest that we Americans do not understand fire as well as we should.

Fire Interest to Firestarting to Firesetting

Children who start fires are ordinary children in need of education, attention and care. Most children learn about fire through age-appropriate education from parents, teachers and fire prevention programs, but some children suffer from emotional disorders, family dysfunction, and chronic stress that twist natural interest into dangerous behaviors.

Fire involvement—the ways in which children experience fire—tends to follow three stages: fire interest, firestarting, and firesetting.

Most children experience fire interest between the ages of three and five as they learn about their environment and experiment with how to control those surroundings. That interest often takes the shape of questions and play, such as wearing fire helmets, playing with toy fire trucks and pretending to cook food on toy stoves. Such interest in fire is a hint for parents to start teaching their children that fire is a tool rather than a toy, that fire is hot, that hot things burn, and burns hurt.

Children who are in the firestarting stage of involvement experiment with ignition sources such as matches, lighters and magnifying glasses. They're still investigating how the world works and how they fit within it. With supervision, such as lighting candles on a birthday cake or monitoring a campfire, these children learn appropriate and safe ways to interact with fire. However, many children also have at least one unsupervised firestart that is usually motivated by curiosity. They use available ignition sources and available fuels, and either extinguish the fire themselves or call for help because the resulting fire was unintentional.

Without educational intervention, for both children and parents, starting unsupervised fires may become a pattern with increasing odds of igniting a destructive fire. During this stage, most children do get the message that fire is dangerous; they learn proper behaviors from their role models and fire prevention programs, but some children do not.

Children who seek ignition sources and collect fuels (and accelerants) are in the firesetting stage. They are no longer curious about fire; they use fire to express themselves and their needs. Their behavior may be motivated by psychological or social disorders, anger, revenge, or desire for attention. If fire eases anxiety and fear, they will ignite additional fires. They rarely attempt to extinguish their fires preferring instead to watch the destruction happen and vicariously feel relief.

Cognitive Development Stages

Understanding youth firesetting requires a short exploration into the cognitive development of children, and the work of Jean Piaget is the best route for that journey. Working with patients and his own children, Piaget separated cognitive development into four stages: Sensorimotor, Preoperational, Concrete Operational and Formal Operational. Each one sheds light on how we humans build relationships with fire.

The first stage—**Sensorimotor**—covers the first two years of life. Parents are taught to child-proof their homes to keep dangerous tools beyond the reach and sight of increasingly mobile toddlers. Hiding matches and lighters in purses or drawers is effective for these children because when an object is out of sight, it no longer exists. That's why the "peek-a-boo" game works and why these kids cry when parents walk out of a room.

Children in the second stage—**Preoperational**—are typically between two and seven years old. They are developing language and an ability to think with symbols. They can think logically in one direction. They are unable to consider perspectives other than their own. Empathy is out of the question. As a result, asking these children to consider how their fire made others feel is ineffective. They can't. (But when I ask fourth graders to consider who might be impacted by a fire and it's fun to see them grasp empathy). Some of these kids also adopt certain topics—dinosaurs, cars, sports, fire—and become experts. They believe that they know more about the topic than anyone else, which makes changing their attitudes and behaviors much more difficult.

These kids also stop falling for the out-of-sight trick. They know that parents hide matches in certain places. Thus parents would be smart to use locking cabinets to keep dangerous tools beyond the reach of their children.

Children in the **Concrete Operational** stage are between the ages of seven and eleven. They believe the world is logically stable, changes can be reversed, and elements can be changed yet retain their original characteristics. Unfortunately, parts of the world don't follow that rule. When liquid propane evaporates, its volume increases 270 times. Similarly, a pool of liquid gasoline might look safe and controlled, but it is surrounded by a large cloud of gasoline vapor that can surprise a match-wielding child. Nor can the changes caused by fire be reversed. Tape doesn't work in this case. These children also thrive on organization and routine. When routines are disrupted, they can experience profound emotional pain.

During the **Formal Operational** stage, which Piaget says typically occurs after the age of 11, children can think abstractly and consider what could happen instead of only what does happen. They can reason deductively and embrace hypothetical situations. That means they can recall experience with small fires and start to experiment with what-if scenarios. They may also take incomplete information, from the Internet for example, and try to replicate or improve the experiment based on their own reasoning. As adolescents experience the neurological growth spurt that slows the parts of the brain responsible for emotional control, recalling social rules, and considering consequences, they may put themselves in risky situations without grasping risk the way they would as adults.

Helping Children and Their Families

Growing up in a small town, I remember when the neighborhood troublemaker was playing with matches and bottle rockets and ignited the grassy embankments on both sides of Clear Creek. Elderly neighbors carried buckets of water to the fire on their side of the creek; firefighters extinguished the fire on the other side before it reached a large propane tank and adjacent gas station.

Children and adolescents involved in unsupervised firestarting and intentional firesetting need assistance. Simply touring a fire station isn't sufficient or even useful; such a fun reward provides positive reinforcement to a negative, dangerous behavior. The appropriate response is a juvenile firesetter intervention assessment. Using the language of behavior modification, enrollment in such a program is a positive punishment because it adds something to the child's life in order to decrease a behavior.

A juvenile firesetter program identifies the level of risk a child has for future inappropriate fire involvement and identifies intervention options. We utilize evidence-based survey tools to determine that risk level and the next appropriate resources in a chain of intervention. Although children are the focus of the program, parents must be involved because most firestarters and firesetters are reacting to other stresses in their lives. Individual healing requires family healing. Part of the intervention enables us to refer families to other experts, from the family doctor to clinical psychologists, in order to continue healing and learning from the social context of their firesetting.

I would like to see mandatory youth firesetting prevention and intervention programs in all communities. Such programs can help identify children who are starting fires before those fires injure or kill. The kids I typically assess are preteens and teenagers who know the basics and are ready to experiment to see what else can happen:

13-year-old boy: *"He showed me a tin can that was full of fire-starting materials. Next I gave him the lighter when he asked for it and he wanted to show me what it would do when he lit it. First, he lit a piece of dryer cloth and blew it out after it burned for a second. Next he lit a paper towel that was soaked with Windex and other aerosols [sic]. Unlike the dryer cloth he couldn't blow it out. Next he dropped it on the sleeping bag and we were able to pat it out."*

12-year-old boy: *"I was really bored so I decided to make a fire. I made the fire by getting matches from a cabinet in the laundry room. When I was at the forest I found some dried leaves so I gathered [them] all up into a big pile. When they were compressed together I lit it. I was thinking that I was cool when I did it because I thought no one else could do it or know how to do it."*

15-year-old girl: *"I found a piece of wood. [He] poured gas all over it and lit it. I didn't really care when the police and fire people came."*

17-year-old boy: *"He pulled out a lighter and set a toy on fire. At first I just stood by and watched, but as I watched the plastic started to melt and burn into the coolest colors."*

Many kids set fires to express trauma and other crises in their lives: divorce, home foreclosure, loneliness, parent military deployment, and sexual abuse, among others. Although they need a level of care I'm not qualified to provide, I take heart knowing I'm helping their families identify community resources that might be able to deliver for them. I am qualified and comfortable educating children and adults who are ignorant of fire and its power, who need an instruction manual about fire or who need a refresher course.

A few years ago, we assessed a boy who had endured several years of abuse from his biological father. His mother didn't contact the police department and fire department after he started burning small pieces of paper, dead leaves and insects. She contacted us after she found him researching do-it-yourself explosives on the Internet. During our assessment, he only admitted to a handful of fires and he said he started them because he was pretending his father was in the flames. He was that angry. The boy didn't offer why he was investigating bomb-making, but I wonder if his anger was reaching into a new realm.

Although I empathize with his anger, sense of betrayal, depression, loss and other emotions that I thankfully haven't experienced, the problem with using fire to vent those emotions is that fire burns indiscriminately. It doesn't internalize what the fire-maker attributes to it. Fires lit by this scared, angry boy could burn anything in its path. Fire burns. And most consequences are borne by the fire-maker.

I can imagine this boy going home and renewing his practice of starting fires because a juvenile firesetter intervention only evaluates the risk of future fire-setting so that trained therapists can start addressing the causes of that behavior. Tomorrow, he uses a lighter to ignite body spray and dead leaves in an aluminum-can "stove" to destroy another mental image of his dad. This time a burning leaf escapes the can when a gust of wind sweeps into the garage. The leaf drifts into the recycling bin and quickly ignites a pile of newspapers. Fire spreads quickly up the wall and across the work bench while he runs outside and starts yelling for help. A neighbor calls 911. Firefighters extinguish the fire before it extends beyond the garage, which is a complete loss. I can imagine worse scenarios, too.

Investigators quickly link witness statements with physical evidence and their prior contact with the family to determine this boy started the fire despite his claims to the contrary. Because he is older than 10, the boy faces charges of felony arson and he is wholly responsible for the monetary costs of the fire. Rebuilding the garage, paying the hotel bill, and paying for their landlord's emotional pain and suffering totals several thousand dollars, which takes a large bite from his mother's wages from the restaurant where she works to support him and his younger sisters.

Felony Arson Charges Not Just for Adults

In Colorado, children who are at least 10 years old are deemed "culpable" for their decisions regarding fire. They are responsible for the consequences of their actions. Our agency, like many others around the country, refuses to reward children who set fires with tours of fire trucks. Instead, we enable our investigators to charge them with arson, which is a felony in our state, because those charges are often the only way to get families of youth to pursue the help they need to address all the other destruction in their collective lives.

That philosophy isn't necessarily popular. When a couple of boys lit a piece of paper on fire and tossed it into some shrubbery, it quickly ignited and scorched the wall of a townhouse in Aurora. The resulting debate focused on how the boys were treated by the district attorney's office. "Unfortunately, Jacob was being a ten-year-old boy," his mother told a reporter in January 2011. Jacob's recollections appeared in *Westword*:

"We were just walking around, being bored," Jacob says. "[The older boy] told me to light this piece of paper on fire. I was like, okay, but I wasn't lighting anything else. It burned him, and he threw it into the bush. And I tried to get it out, and I got burned. Then we started throwing rocks at it, trying to put it out."

The bush in question was one of several dry evergreens that lined one side of a townhome a couple of blocks from Jacob's place. It went up like last year's Christmas tree.

"It started catching on the center bush," Jacob recalls. "Then it lit the first bush. And it went to the third and fourth and then to the house. It just happened. We both started running, and we were yelling for help. Somebody saw the fire and called the cops, and the fire department people came up. I'd never seen anything like it except on the news."

The district attorney's office pursued second-degree arson charges, much to the chagrin of Jacob's mother, her attorneys and the media who rushed to their defense.

"That's what boys do," an attorney told the reporter. "They play with magnifying glasses and matches, they have rock fights. I don't think you can say that Jacob set out with the intention to burn a house down." She believed civil action to recover the $195,000 in property loss was more appropriate than criminal action. From our perspective in the fire service, financial losses are important, but personal responsibility and life safety concerns trump property loss.

Alan Prendergast, the reporter for *Westword*, suggested a misdemeanor charge was better than a second-degree felony because the boys allegedly had no intention of causing the damage.

Misusing fire isn't an insignificant action. It's a serious problem with potentially serious consequences, regardless of intention. I doubt most drunk drivers intend to kill others, but they do, and prosecutors, juries and judges use felony charges to punish them. Where is the outcry in those cases? If those kids had been playing with a gun that day, and unintentionally shot into a townhome, the community would have demanded harsh justice. I'm stumped why the reaction differs when a child uses fire—an equally potentially violent tool—to express personal issues ranging from boredom to anxiety to anger.

As noted by US Forest Service researcher Louis Barrett in the opening quotation, the Me-Wuks of what is now California limited burning to specific individuals because they deemed fire use a significant responsibility. Fire destroys. Learning how to use it safely and wisely is essential. We need to fill gaps in children's and sometimes parents' education about fire—possibly even educating them how to kindle and use fire safely. We take that strategy with hunter education, driver education and sex education, but not with fire safety education.

Psychologist Lev Vygotsky postulated that children learn within zones of proximal development. When they are unable to solve a problem, they can succeed with guidance from someone with more advanced skills. Ultimately, it's the idea of mentoring. Additionally, psychologist Kurt Fischer has expanded Piaget's theories, suggesting a supportive social context is the most important factor in cognitive and skill development. It might be time for fire departments to teach safe fire use—true fire safety—rather than outright fire prevention.

Last year an elementary school teacher told me a story of her college-aged daughter who received burns from fireworks because she didn't know how to use them safely.

This past 4th of July she was asked to join [her boyfriend's] family. She went and had a 4th of July surprise. In Kansas and Nebraska, fireworks are not illegal; anyone can buy and light off fireworks. Her boyfriend took her to buy some fireworks and they planned to light them off at dark. She had never participated in lighting fireworks. Her boyfriend tried to show her how to light one and her shirt caught fire. He went to get another shirt and she burned a hole in that shirt as well.

She did not grow up lighting them and she had no knowledge of how to handle a firework correctly. It is a funny story the way they tell it, but are Coloradoans doing a disservice to kids by not teaching them how to handle fireworks?

Her parents had never used fireworks and thus had never modeled appropriate behavior. Finding herself in a zone of proximal development, she turned to her boyfriend to learn how to use fireworks but was burned in the process. A class in fireworks safety—with time for practice—instructed by subject matter experts may have prevented her burn injury.

A recent press release from the American Pyrotechnics Association suggests that such a tactic—increasing education while lowering enforcement may be working. It claims that while fireworks consumption in the United States has increased from 41 million pounds in 1980 to over 207 million pounds in 2012, the injury rate has declined dramatically. Although the rate has declined, the number of injuries that result in emergency room visits has remained constant, according to the NFPA, between 8,500 and 9,800.

Ignorance Does Have Consequences

Many children get burned by their fires simply from ignorance. A fourth-grade student in one of my classes wrote a thank you letter to me and included the following admission: "When I learned about how big of a fire a kid could make and the consequences that the kid could have, I started to think differently. It amazes me that one small fire can start a huge fire." That child was grappling with the Piagetian idea of conservation in which a fire's growth shattered his expectation that a small amount of fuel would create a small fire. Another wrote, "I would have never known that starting a fire could mess up your entire life." That child had started understanding abstract reasoning. Had he been younger, that type of understanding would be impossible.

Some communities are blessed with better resources than others to address youth fire misuse. Plenty of programs exist that not only assess youth and their families for future firesetting risk (using forms such as the ones in the Appendix), but also empower kids and families to limit access to firesetting supplies, improve communication within families, provide access to community resources for at-risk families, and create better alternatives for expressing fear, anger and depression.

Collectively it's time we recalibrate our understanding of fire because it is a powerful phenomenon. It is sacred and profane. It creates and destroys. It is powerful, yet it is not magical. It follows the rules of physical science and chemistry as we know them. ⤝

4

Fire's Dark Side:
A Tool of Pain and Destruction

The beasts of the field cry also unto thee: for the rivers of waters are dried up, and the fire hath devoured the pastures of the wilderness.

Joel 1:20, King James Version of *The Bible*

We suffered very much in this passage; for the Savages having set the Herbs of the Plain on fire, the wild Bulls were fled away, and so we could kill but one, and some Turkey-Cocks.

Louis Hennepin, *A New Discovery of a Vast Country in America*

*A*s a tool, fire has transformative powers. In untrained, careless or malicious hands, it is dangerous, even deadly. Ancient and modern people have used fire to destroy resources, from animal forage to infrastructure in formal warfare, as we read in the opening quotation from French explorer Louis Hennepin recorded in 1680 near the headwaters of the Illinois River. None of these changes can be undone. The chemical reaction of combustion is permanent. The finality of fire's impact on a given fuel is hard for many people to grasp, especially children.

Fire transforms buildings, communities and ecosystems. Based on data from the National Fire Incident Reporting System (NFIRS), the National Fire Protection Association has re-interpreted the cost of fire within the United States. Instead of relying on property-loss figures, its experts took an economic perspective and included insurance coverage, the cost of operating career fire departments, new building costs for fire protection, the monetary value of donated time from volunteer firefighters, and a monetary equivalent for civilian and firefighter deaths and injuries sustained during fires. In other words, they combined loss figures with costs of prevention, mitigation and suppression efforts. For 2011, that cost was $329 billion, approximately two percent of the nation's gross domestic product.

Fire ecologist Robert Gray is performing similar calculations for wildfires, moving beyond the costs of suppression and short-term land rehabilitation to estimate the economic impact of wildfires. Alberta's Slave Lake Fire in 2011, for example, destroyed 40 percent of the town including residential and commercial zones. Those costs are easy to determine based on insurance claims. However, many of the people who lost their homes were medical professionals who moved away from the community and took their skills with them. The costs to the community in lost productivity, lost wages and untreated medical conditions—including all the children and adults who began experiencing post-traumatic stress conditions as long as a year after the fire—continues to accumulate. Wildfires also impact tourism, timber harvests, infrastructure, utility usage, and property values that in turn impact governments that rely on property taxes and watersheds. Erosion following the 1996 Buffalo Creek Fire dumped unprecedented amounts of sediment into Strontia Springs Reservoir, which supplies 75 percent of the drinking water needed by the City of Denver. Those costs, both the known and unknown, are staggering.

That Hot, That Quick

Idaho Springs, my hometown, is among a few urban communities in Colorado's mountains that has escaped a town-razing fire in its history, but it certainly has had its share of big structure fires, including one that had monstrous potential and left a mark in my memories of childhood. The Sampler Mill, an architectural relic from the town's 19th-century gold-mining roots, stood until February 26, 1980.

That morning several kids had broken into the mill as they had done numerous times before. This time, they brought gasoline or other petroleum products and some matches. Before the boys' match hit the puddle of gasoline on the building's wood floor, the flaming tip encountered a cloud of gasoline vapor hovering in the air. The vapor cloud exploded violently spreading flames across the floor and around the room's aging, musty walls, which provided easy fuel for the hungry flames. As fire ripped through the building, it consumed old machinery, automobile parts and other junk.

The town's civil defense sirens began blaring at 8:02 a.m. Volunteer firefighters, many of whom noticed the smoke in the valley as soon as they stepped outside to get into their personal vehicles, responded. The first truck arrived three minutes after the sirens started sounding, according to Fire Chief Jim Albers. He immediately called for mutual aid from Dumont and Georgetown, communities four and twelve miles to the west.

"Jerry and I got in Number One," firefighter Dean St. John, one of my mentors in the Idaho Springs Fire Department, recalled for me. That 1977 Chevrolet truck had 250 gallons of water and a small 70-gallon-per-minute pump. It was small for a structural fire, but most of the agency's apparatus were staffed for this incident.

We were the first two there. He gave me water, I went in the walkway and sprayed that thing but the fire just went up over my head. I don't know what was in there, but I just grabbed all the hose up and ran, saying 'get the hell out of here.' He took off and I was running up the sidewalk carrying the hose.

They moved the truck, but not before the fire's heat seared its paint and melted the light bar. "It was that hot that quick," St. John said. "From there on it was downhill."

As the Sampler Fire swelled into an inferno, radiant heat rapidly dried and warmed the wood siding beyond its ignition point, wind pushed superheated air onto adjacent buildings and flaming embers dropped onto dry fuels. Flames seemingly jumped to a pair of two-story houses east of the mill and onto a house to the north. Within 30 minutes, 10 engines and 25 volunteer firefighters were battling the conflagration, working to contain the blaze to the single block, while the wind threatened to expand those boundaries.

Containment was imperative. Homes surrounded the mill to the north and west. An automotive repair shop was across the street to the east, and less than a block downwind was a trio of gas stations that welcomed motorists from I-70 into the downtown historic district of Idaho Springs. One block downwind on the north side of the street was another hazard: Idaho Springs Elementary School, which was in session that chilly February morning.

"It was so hot, we couldn't get close to it," firefighter Jimmie Cullen recalled. His crew stretched a pair of hoses from a hydrant-fed engine to create a water curtain between the burning mill and the A&B. Having witnessed a similar fire obliterate another mill and threaten the town back in 1956, Cullen knew the odds were against the firefighters. "We sprayed water like crazy and it was just evaporating," firefighter Bob Jones recalled. "It never hit the buildings." Without water absorbing the intense heat, the conflagration would have spread at will.

The wind, which provided a continually replenishing supply of oxygen, carried embers as far to the east as City Park—ten blocks away—but diminished as the morning lengthened. "If the wind had kept blowing, we could have lost this town," one firefighter told a reporter.

In all, the fire caused thousands of dollars in damage to buildings and their contents, although the amount of damage to the property owners' and townspeople's sense of security was immeasurable. In the weeks that followed, a retired firefighter demolished the old Ruth Mill on his property to prevent it from experiencing a similar fiery and dangerous fate. He had learned an important lesson about fire and its potential consequences.

I was in second grade when this fire occurred. It informed my imagined version of other big city fires like the Great Chicago Fire that inspired the creation of Fire Prevention Week.

Chicago, 1871

In 1871 the city of Chicago was a big city in the otherwise rural West. It also was a city of contrasts, as Peter Hoffer explains in his book *Seven Fires*, where palatial homes of corporate barons were around the corner from row houses filled with immigrant laborers and streets crawling with legions of homeless. One common thread through the city was its flammability: lumber in the yards and wheat in the grain elevators; miles of wooden sidewalks and planked thoroughfares; endless streets of simple clapboard houses and avenues of elegant stores and mansions.

Chicago also featured a new type of wood-frame house. The balloon-frame, invented in the city in the 1840s, replaced heavy timber beams with an open arrangement of milled four-by-four and two-by-four boards. Sheets of thin wood were nailed to the lightweight framing to finish walls and roofing. Then, as now, balloon-framing allowed builders to construct housing more quickly and less expensively. It also increases the surface area-to-volume ratio of wood, which lowers the amount of heat needed to ignite the lumber.

The Great Chicago Fire ignited on October 8, 1871, in a barn owned by Patrick and Catherine O'Leary probably after a couple vagrants dropped their smoking materials in the hay. Rather than stomp out the fire, they ran to the house to report seeing flames in the barn. Heat from the smoldering sticks of paper and tobacco found plenty of hay and straw on the floor to burn. As the small fire grew, it found two tons of hay and two tons of coal within the barn, as well as the dry structure itself, and plenty of oxygen as a cold front moved over the region producing gusty southwesterly winds. Flames quickly jumped from the barn onto adjacent structures, produced a convective column capable of generating its own wind and hurled embers hundreds of feet downwind.

Aurelia King, the wife of a prominent clothing merchant, recorded the fire's growth in a letter to friends:

> *At one o'clock we were wakened by shouts of people in the streets declaring the city was on fire—but then the fire was far away on the south side of the river. Mr. King went quite leisurely over town, but soon hurried back with the news that the courthouse, Sherman House, post office, Tremont House, and all the rest of the business portion of the city was in flames, and thought he would go back and keep an eye on his store.*
>
> *He had scarcely been gone fifteen minutes when I saw him rushing back with his porters, bringing the books and papers from the store, with news that everything was burning, that the bridges were on fire, and the North Side was in danger. From that moment the flames ran in our direction, coming faster than a man could run. The rapidity was almost incredible, the wind blew a hurricane, the air was full of burning boards and shingles flying in every direction, and falling everywhere around us. It was all so sudden we did not realize our danger until we saw our Water Works (which were beyond us) were*

burning, when we gave up all hope, knowing that the water supply must soon be cut off.

Chicago's firefighters responded, but they did so as exhausted foot soldiers having battled a fire that destroyed four blocks the previous night. Their equipment needed repairs, their horses needed rest and they needed time to recover physically and mentally. Instead they found themselves back in the trenches facing an unbeatable enemy.

Chicago attorney Jonas Hutchinson wrote a letter to his mother on October 9, 1871, describing the fire as it chewed through the city:

What a sight: a sea of fire, the heavens all ablaze, the air filled with burning embers, the wind blowing fiercely and tossing fire brands in all directions, thousands upon thousands of people rushing frantically about, burned out of shelter, without food, the rich of yesterday poor today, destruction everywhere.

The official death toll was 200, but many of Chicago's residents were undocumented, having arrived recently from overseas and other parts of the nation. Hundreds of residents also lived in the city's shadows beyond the reach of official tabulators and services. Other costs include $200 million of real property loss, which is over $3 billion in today's dollars. Over 90,000 people became homeless. Flames destroyed wheat, corn, meat, coal and lumber stockpiles. It devastated the city.

Fire Changes Us

As the following ancient stories reveal, the consequences of fire can explain why certain critters look the way they do. For my daughters who are still young enough that fantasy and reality mix easily, these stories help them understand long-term consequences of fire misuse, in part, because they know many of the animals. They see red-breasted robins in our backyard and recall ostriches from our safari ride at an amusement park. The lessons here could create a safer foundation for their future contact with—and even experimentation with—fire.

Peer deeper into the tales and imagine your reaction to a fellow robin who burned her chest and changed the species forever, or the ostrich, still unwilling to fly, who seems forever ashamed of losing his treasured fire. Ultimately I wonder how our behaviors and decisions would change if our fire use altered our own species forever. Maybe it does.

The San people of Africa's Kalahari Desert survived for thousands of years in a harsh environment.

> One day, Mantis smelled a wonderful aroma floating through the countryside. Curious, Mantis peeked through a bush and saw Ostrich roasting food over a fire.
>
> When Ostrich finished eating, he took the fire and tucked it under his wing. Mantis had never seen fire, and he now wanted it for himself.
>
> When Ostrich jogged by, Mantis called out to him and told him of a wonderful tree filled with fruit. Excited, Ostrich followed Mantis to a tree covered with yellow plums.
>
> "The best fruit," said Mantis, "is at the top."
>
> Ostrich eagerly reached up with his long neck and extended his wings to keep his balance. As soon as Ostrich opened his wings, Mantis snatched the fire and fled. Since then, Ostrich has kept its wings at its side and has never attempted to fly.

This story fits their egalitarian culture, punishing Ostrich with eternal shame for not sharing.

Fire's consequences are greater than dollars wasted when property burns or when medical bills arrive at the insurance company or in the mailbox. The costs of fire can be physical, emotional and psychological: injuries, scar tissue, changed physical appearance, changed life opportunities, different outlooks and new priorities.

Consequences can be devastating when it comes to stealing such a powerful and prized possession from skookums, gods or other creatures. Sometimes the being who excluded others from fire pays the price, as with the ostrich, and as exemplified by this excerpt from a Maidu story from what is now northern California:

> At one time the people had found fire, and were going to use it, but Thunder wanted to take it away from them, as he desired to be the only one who should have fire. He thought that if he could do this, he would be able to kill all the people.
>
> After a time he succeeded and carried the fire home with

him, far to the south. He put Woswosim (a small bird) to guard the fire so that no one would steal it. Thunder thought that people would die after he had stolen their fire, for they would not be able to cook their food, but the people managed to get along.

Mouse and other people went to retrieve the fire. They took a flute with them for they meant to put the fire in it. They traveled a long time, and finally reached the place where the fire was.

After a while Mouse was sent up to try and see if he could get in. He crept up slowly till he got close to Woswosim, and then saw that his eyes were shut. When Mouse saw that the watcher was asleep, he crawled to the opening and went in. This done, Mouse took the flute, filled it with fire, then crept out, and rejoined the other people who were waiting outside.

For a while all went well, but when they were about halfway home, Thunder woke up, suspected that something was wrong, and asked, "What is the matter with my fire?"

He jumped up with a roar; his daughters were thus awakened and also jumped up. They carried with them a heavy wind and a great rain and a hailstorm, so that they might put out any fire the people had. Thunder and his daughters hurried along, and soon caught up with the thieves, when Skunk shot at Thunder and killed him.

Then Skunk called out to the daughters, "After this you must never try to follow and kill people. You must stay up in the sky, and be the thunder. That is what you will be." The daughters of Thunder did not follow any farther; so the people went on safely, and got home with their fire, and people have had it ever since.

Sometimes, thieves pay the price, and occasionally the price of fire is death, as in the following three stories from the Tolowa of Southwestern Oregon, Nimipu of the Columbia River Plateau and Aka-Jeru of the Andaman Islands in the Indian Ocean.

The Indians had no fire. The flood had put out all the fires in the world. They looked at the moon and wished they could secure fire from it. Then the Spider Indians and the Snake In-

dians formed a plan to steal fire. The Spiders wove a very light balloon, and fastened it by a long rope to the Earth. Then they climbed into the balloon and started for the moon. But the Indians of the Moon were suspicious of the Earth Indians, so the Spiders said, "We came to gamble." The Moon Indians were much pleased and all the Spider Indians began to gamble with them. They sat by the fire.

Then the Snake Indians sent a man to climb up the long rope from the Earth to the moon. He climbed the rope, and darted through the fire before the Moon Indians understood what he had done. Then he slid down the rope to Earth again. As soon as he touched the Earth he traveled over the rocks, the trees, and the dry sticks lying upon the ground, giving fire to each. Everything he touched contained fire. So the world became bright again, as it was before the flood.

When the Spider Indians came down to Earth again, they were immediately put to death, for the tribes were afraid the Moon Indians might want revenge.

* * *

Long ago, the Nimipu had no fire. They could see fire in the sky sometimes, but it belonged to the Great Power. He kept it in great black bags in the sky. When the bags bumped into each other, there was a crashing, tearing sound, and through the hole that was made fire sparkled.

People longed to get it. They ate fish and meat raw as the animals do. They ate roots and berries raw as the bears do. The women grieved when they saw their little ones shivering and blue with cold. The medicine men beat on their drums in their efforts to bring fire down from the sky, but no fire came.

At last a boy just beyond the age for the sacred vigil said that he would get the fire. People laughed at him. The medicine men angrily complained, "Do you think that you can do what we are not able to do?"

But the boy went on and made his plans. The first time that he saw the black fire bags drifting in the sky, he got ready.

First he bathed, brushing himself with fir branches until he was entirely clean and was fragrant with the smell of fir. He looked very handsome.

With the inside bark of cedar he wrapped an arrowhead and placed it beside his best and largest bow. On the ground he placed a beautiful white shell that he often wore around his neck. Then he asked his guardian spirit to help him reach the cloud with his arrow.

All the people stood watching. The medicine men said among themselves, "Let us have him killed, lest he make the Great Power angry."

But the people said, "Let him alone. Perhaps he can bring the fire down. If he does not, then we can kill him."

The boy waited until he saw that the largest fire bag was over his head, growling and rumbling. Then he raised his bow and shot the arrow straight upward. Suddenly, all the people heard a tremendous crash, and they saw a flash of fire in the sky. Then the burning arrow, like a falling star, came hurtling down among them. It struck the boy's white shell and there made a small flame.

Shouting with joy, the people rushed forward. They lighted sticks and dry bark and hurried to their tipis to start fires with them. Children and old people ran around, laughing and singing.

When the excitement had died down, people asked about the boy. But he was nowhere to be seen. On the ground lay his shell, burned so that it showed the fire colors. Near it lay the boy's bow. People tried to shoot with it, but not even the strongest man could bend it.

The boy was never seen again. But his abalone shell is still beautiful, still touched with the colors of flame. And the fire he brought from the black bag is still in the center of each tipi, the blessing of every home.

* * *

In the days of the ancestors they had no fire. Biliku had fire.

While Biliku slept, Maia Lirčitmo (Sir Kingfisher) came and stole fire. As he was taking the fire Biliku awoke and saw him. Lirčitmo swallowed the fire. Biliku took a pearl shell and threw it at Lirčitmo and cut off his head. The fire came out (of his neck). The ancestors got the fire. Lirčitmo became a bird.

While those stories consider the acquisition of fire, they don't consider how humans deal with fire and, in the process, are also transformed. Many of us in the fire service know how fire transforms us personally. Warren Yahr wrote about his experiences as a lookout for the U.S. Forest Service to share those transformations. The following tale of his first wildfire is an excerpt from his book *Smokechaser*:

"How's she look?"

"Can't see much change. It's still blazing up pretty good, but not getting any bigger."

"Well, that's something. Let me know if there's any change."

"Right."

Another hour went by and I could feel footsteps on the stairs. Suddenly, Bob burst in.

"I gotta call Wade. The tree is just too hot to handle. Every time I chop at it, a bunch of burning stuff falls down on me. Look at my hat."

I could see what he meant. The hat had been burned through in a dozen places. While he explained the situation to Wade, I started getting my boots on, hoping I might get to go. I was almost ready when the conversation ended.

"Wade says we should both go and take a saw along. Once the tree is down, one of us is to come right back here."

"Did he say who?"

"No. We can decide that ourselves."

"Okay, let's go. I'm ready."

"Just as soon as I gas up this lantern. My batteries are just about finished and it's really dark out there."

He was right, of course. The night was pitch black and the wind was still blowing furiously. The burning tree, however, acted like a beacon and as we neared it I could see burning chunks fall off and land in a shower of sparks. Luckily the surrounding area was still wet from the rain so the fire wasn't spreading. That tree would have to come down, though, before things dried out or we'd have real problems. As it was we'd have enough.

The tree was aflame along its entire length. First, we'd have to cool down the lower section before even starting to saw it down. It was too hot to get near otherwise.

"See what I meant," shouted Bob over the roar of the flames. With the wind blowing and fanning the fire, the whole thing was burning like a torch.

"I'll say I do," I shouted back, starting to throw shovelfuls of dirt on the burning trunk.

When it was cool enough to start sawing, we began, one of us keeping an eye overhead for falling debris. We stayed on the windward side so the smaller pieces blew away from us, but every once in a while somebody would shout, "look out!" and a burning chunk would come crashing down as we jumped aside. Smoke came pouring out of the saw cut and burst into flame as we cut deeper, and we had to stop and throw on more dirt to cool things down. At last the cut started to open up and Bob yelled timber and over it went in a shower of sparks and flaming chunks.

"Boy, what a sight," grinned Bob.

"You said it. I'll stay and put it out if you like."

"You want to?"

"Sure. I could use the practice."

"Well, okay, if you're sure that's what you want."

"Yup, just as long as you can be trusted alone with those pies."

"Heck, I'd forgotten all about 'em. I promise they'll be all there when you get back."

"Oh, go have a piece. You gotta be hungry."

"I am, but a promise is a promise."

"Whatever you say. Just take off now so I can get to work."

There was no way he could find his way back without the lantern, so I was left to work by what light there might be from the fire. Throwing dirt to cool things down consumed a couple hours, and when no flames remained I was left in total darkness.

Stumbling around I located a spot out of the wind that was fairly dry and I laid down to rest. I could do no more until daylight came...

When the first rays of sunlight appeared I took off my gloves and, as we had been taught in fire school, felt every square inch of tree trunk to make certain no hot spots remained. Satisfied, I dug a trench around it and that was that. A final once over, and I shouldered the fire pack and started for home.

Legacies of Fire

Yahr matured as he worked that fire, a feeling many of us in the fire service have experienced with a first or unique fire. A fire deliberately set to burn a large Idaho Springs business in February 2003 was a watershed moment in my firefighting career.

Having a gentle slumber rudely awakened by the piercing chirping of my pager wasn't rare; in fact, it happened more often than not. On that February night in 2003, however, the words from the dispatcher sounded direr than for other night calls. This time, she alerted East and West Fire to a possible structure fire in the 1700 block of Colorado Boulevard in the core of Idaho Springs. Anytime a structure fire was possible, I tried to respond because the need for staffing would be great, especially on a weekday night such as this one.

I rolled out of bed and looked west from the bedroom window of my condo. I didn't see anything suspicious then; I reassured myself that I'd no doubt be back soon. I hurried downstairs, dressing along the way, grabbing my keys, locking the door behind me and turning my Jeep Cherokee around to head west along Riverside Drive to the fire house. Approaching the 23rd Avenue Bridge, I noticed a glow to the west—not a good sign. I also noticed that the first page had not netted many firefighters; three of us were from the east end of town, suggesting we and the other firefighter who lived two houses away from the station would be the main crew for some time.

We donned our gear and hurried to the newest engine in our fleet, a hand-me-down from Aurora, Colorado, when it had upgraded to new.

Jason engineered as he always did; as a lieutenant I took the officer's seat and two firefighters, including another lieutenant, climbed into the back seats and prepared their SCBA bottles.

In those few minutes, the Idaho Springs Police sergeant had responded and found a fire at the northeast corner of the Gold Rush Center, a two-story wood-frame strip mall with four suites on each level. We left the station, lights and siren blazing, and turned south up 20th Street, then west on Miner Street to connect to a hydrant at Miner and Feistner Boulevard. Jason parked and began helping Chris, a relatively new firefighter, and Rick, the fellow lieutenant, connect a five-inch hose to the hydrant while I started walking toward the property, strapping on my SCBA backpack as I crossed the Colorado Boulevard Bridge over Clear Creek.

The fire had started at the northeast corner in the chiropractor's office on the second floor. I radioed the Sheriff's Office in Georgetown for a second page for a confirmed structure fire and established myself as Gold Rush Command. It was my first large assignment in the role of incident commander. As I walked from the east end of the building to the west end, charged with adrenaline, the fire sprinted through the open attic—unchallenged by firewalls that should have existed to block such fire growth—and won the race to the west wall. It had vented through the eaves in the seconds it took me to reach the western parking lot. I updated dispatch with the fire's status, completed my 360-degree walk around the back of the building from Placer Street, which is above the two-story wood-frame commercial structure, and looked back to the engine to see that it had only now driven from the hydrant to the southeast corner of the property. At least one 2½-inch line was stretched from the hydrant instead of the 5-inch line, which meant that our water supply would be short of my expectations until additional resources could arrive.

I returned to the front of the building and found that the engine was not yet pumping water and my crew of firefighters had yet to stretch hose to the fire. Fortunately, other firefighters had started arriving in that period, bringing a ladder truck to the north side, and crews were en route from Stations 1, 3, 4 and 7 (Dumont, Empire, Georgetown and Alice/St. Mary's). When Rick and Chris did maneuver their 1¾-inch line to the fire and water swelled the double-jacketed hose, they hit the fire with a heavy stream and pushed it into the building from its origin. That I hadn't caught the mistake was a sign of my own adrenaline and my attention to other details of incident command such as directing the incoming units to connect to a hydrant west of the building.

In the end, we held the fire to the second floor and limited water and smoke damage to the businesses in each of the four first-floor occupancies. It wasn't a perfect attack—far from it—but we extinguished the fire safely and preserved enough of the fire's origin that an arson dog later detected traces of hydrocarbons, suggesting the fire had been set deliberately with gasoline, kerosene or another petroleum-based liquid. It was the largest structure fire I had been on to that point and the largest I have been on since; it was a milestone in my education and experience as a volunteer firefighter.

That fire altered my career in the fire service, much as the burning tree impacted Warren Yahr, but there is another side of this coin. The complement to my experience as firefighter is the experience of victims. As a child, I experienced two chimney fires at our house. Both were extinguished by our local firefighters quickly. The only damage to our house—and it was more inconvenience than damage—was some muddy footprints and a mouse carcass left behind by firefighters' boots. A bit of dirt on a carpet pales in comparison to the psychological and emotional trauma that fire can generate.

When fire does burn a vehicle, house or business, we often focus on physical injuries. As long as no one was hurt, everything will be fine because things can be replaced. Furniture, toys, books and other personal property can be replaced. Yet many people who experience fires from the role of victim would disagree. My colleague Deanna Harrington experienced a fire in her Coal Creek Canyon home when she was a child; she still carries trauma from that incident:

It isn't that years ago on a particular Tuesday afternoon my family simply had a bad day, rather it is something that we live with forever. I have come to hate "Throw Back Thursday" on social media. It is good and fun for so many people, but not for those of us who don't have pictures dating back to our childhood years. Recently I was asked to bring my baby picture for a particular event I was invited to and I simply had to decline the invitation to attend the event altogether. And while I do love golden retrievers, I will never again be able to bring one of them into my family. Most of what was lost that day was merely stuff, but it was our stuff and with that stuff we also buried many memories that the items reminded us of regularly. My family walked away physically unscathed, minus the loss of our beloved dog, but I was forever changed. I still often awaken during the night to make sure that smoke alarms are blinking at me to indicate they are watching over my family. I fret a great deal about escape plans and exits when my children travel without me. I also can't tell you how many times I go back in the house to be absolutely certain that the stove is really turned off and the iron is unplugged before I am on my way to work.

Deanna is among the most fervent life safety educators in Colorado, particularly regarding fire prevention. This incident and its legacy influenced her join the fire service, initially as a firefighter. Now she serves as a deputy fire marshal and life safety educator.

Sheets of Flame

As a destroyer, fire has shaped lives across the globe and throughout time. Back in 1871, fire left a significant mark on the city of Chicago, but the weather system that blew a structure fire into that notorious urban conflagration also fueled a massive wildfire in southeastern Wisconsin. The Peshtigo wildfire was the product of countless smaller fires, cultural ignorance and atmospheric conditions. Professors Denise Gess and William Lutz described the conditions of southeastern Wisconsin in October 1871 as hyper-hazardous:

Peshtigo and Marinette were fast becoming unchecked ovens. The ground surface was heating up. Ionized atoms of charcoal combined with trace methane from the dry marshes; gases blew into the upward flowing drafts of relentless combustion that gathered into clouds in the air overhead.

Once ignited, the fire burned over 2,400 square miles, incinerated towns and killed thousands. It took Increase Lapham, a local meteorologist, until August 1872 to file his report and theory on the causes and phenomena of the Peshtigo fire:

> *The explanation already given of the "traveling sheets of flame" is the correct one. Burning gas (carbureted hydrogen) was produced by the excessive heat of the fire much faster than it could be consumed; hence it arose in great masses, taking the place of the atmospheric air; these masses were driven about by the wind, and would cause death by suffocation, precisely as when common gas is allowed to fill a sleeping apartment. Such masses of combustible gas could only be consumed at the surface where they come in contact with the oxygen of the air; hence they would present the appearance of great balls of fire. Whenever the air penetrated the gas, it at once became explosive...there is no evidence either from telegraph wires or otherwise, of any unusual electrical disturbances during the great fires; I think all the facts can be accounted for without calling in the aid of electricity.*

He was close. The clouds of combustible vapor needed oxygen and a heat source—in this case sparks, flames or smoldering peat fires provided the minimum ignition energy. When the clouds ignited they would have looked like fireballs as they incinerated the landscape. His description suggests flashovers, but the accepted definition of flashover is limited to enclosed spaces such as in buildings. Whether inversions or other weather conditions could create equivalent enclosed spaces outside remains an intriguing topic for research yet certainly seems reasonable.

Protecting the Forests from Flames

Forest fires were a perceived scourge of the nation then as now, threatening water supplies and destroying valuable commodities. In 1897, foresters schooled in the principles of scientific management penned a report for the Secretary of the Interior that called for a new policy regarding the nation's forests. They believed all but a handful of wildfires were caused by humans and therefore could be prevented:

Fires are particularly destructive to the forests of western North America. These are composed almost exclusively of highly resinous trees, which, when they grow beyond the influence of the moisture-laden air currents from the Pacific Ocean, ignite easily, and, burning fiercely on the surface, are quickly killed, while the flames sweep forward, leaving standing behind them the dead, although unconsumed, trunks to furnish material for later conflagrations and to intensify their heat. The climate, with its unequally distributed rainfall and intensely hot and dry summers and the peculiarly inflammable character of the forests, make forest fires in the West numerous and particularly destructive, and no other part of the country has suffered so seriously from this cause.

Fires in Western forests are started by careless or ignorant hunters and campers, who often leave their campfires burning or, in utter wantonness, ignite coniferous trees to enjoy the excitement of the conflagration. They can be occasionally traced to the effects of lightning, which locally is held responsible for many forest fires, although in reality fires set in this way are rare, as lightning is usually accompanied or followed by copious rains, which extinguish them before they can gain headway; and very rarely they are produced by the rubbing together of adjacent trees swayed by the wind. The right of way of every railroad crossing the Rocky Mountains and the other interior ranges of the continent is marked by broad zones of devastation due to fires which have started from the camps of construction gangs or the sparks of locomotives; and thousands of acres of timber are destroyed annually by the spread of fires lighted by settlers to clear their farms.

Prospectors in search of valuable minerals frequently set fires in wooded regions to uncover the rocks and facilitate their operations; and the shepherds who drive their flocks to pasture during the summer mouths in the mountain forests of Oregon and California make fires in the autumn to clear the ground and improve the growth of forage plants the following year. No statistics show the area of forests destroyed annually by fires in the Western States and Territories, but nearly every sum-

mer their smoke obscures for months the sight of the sun over hundreds of square miles, and last summer your committee, traveling for six weeks through northern Montana, Idaho, and Washington, and through western Washington and Oregon, were almost constantly enveloped in the smoke of forest fires.

Such conflagrations have occurred in the West since it was settled, and they will always menace the prosperity of that part of the country. Once fully under way, a fire in a forest of coniferous trees will spread until it is extinguished by rain or encounters some natural barrier like a river, or, until driven back over its own course by a change of wind, it expires from want of material. No human agency can stop a Western forest fire when it has once obtained real headway, and the only hope of averting the enormous losses which the country suffers every year from this cause is in preventing fires from starting in the forests or in extinguishing them promptly. They will always occur, but the experience gained in the Yellowstone National Park since it has been patrolled by detachments of the United States Army, and in Canada, shows conclusively that with the aid of disciplined forest rangers intelligently directed the number of forest fires in any district can be greatly reduced, and that it is frequently possible to extinguish small fires if they are energetically attacked when first discovered.

Despite the subject matter experts' urging, humans and the rest of the ecosystem kept creating conditions ripe for wildfire.

Fifty years after Peshtigo and in apparent defiance of the new forest managers' models, a similar combination of factors created an equally tragic scene in the Pacific Northwest. The wildfires of 1910—a series of over 1,700 fires whipped into a fury by extreme weather conditions—transformed the U.S. Forest Service and countless communities in Washington, Idaho and Montana. Although authors Stephen Pyne and Timothy Egan have written definitive histories of the August 1910 firestorms, a description from a participant is useful here. Assistant Forest Supervisor Clarence Swim, who was in Newport, Washington, in August 1910, recalled the firestorm for a USFS report. His recollection notes his own awe and fear as well as complaints about fire's misuse:

The late summer of 1910 approached with ominous, sinister and threatening portents. Dire catastrophe seemed to perme-ate the very atmosphere. Throughout the first weeks of August the sun arose a coppery red ball and passed overhead red and threatening as if announcing an impending disaster. This fiery red sun continued day after day. The air felt close, oppressive and explosive...

August 20 arrived more ominous and threatening than the days preceding. Reports of so many fires came in that it was impossible, with means at hand, to even begin to cope with the situation. The wind began to increase in velocity from the west. Small fires were fanned into large ones. The air was rapidly becoming filled with smoke more dense than previously.

From the window of the office we could see for several miles along the timbered bench lands northeasterly from the river. These yellow pine slopes were occupied by several ranchers. We could see fires break out from these ranch locations and sweep up the slope beyond. It was clearly evident from the lo-cation of these new fires that the ranchers were starting what they thought to be back fires as a protection to their own prop-erty. There is not anything more dangerous than a back fire started by hands of the inexperienced. These fires spread with great rapidity. Finally the expected hurricane broke in all its fury. Local fires burned together and swept through the forest as one vast conflagration.

The stories of Peshtigo and the 1910 Blowup should help us stop and think about current wildfire conditions and the ecological role of humans before, during and after wildfires.

In the 1860s, John Wesley Powell witnessed the full power of fire while exploring mountains in what is now Colorado and wrote about it for *The Century Magazine* in 1889:

More than two decades ago I was camped in a forest of the Rocky Mountains. The night was arched with the gloom of snow-cloud; so I kindled a fire at the trunk of a great pine, and in the chill of the evening gazed at its welcome flame. Soon

I saw it mount, climbing the trunk, crawling out along the branches, igniting the rough bark, kindling the cones, and setting fire to the needles, until in a few minutes the great forest pine was all one pyramid of flame, which illumined a temple in the wilderness domed by a starless night. Sparks and flakes of fire were borne by the wind to other trees, and the forest was ablaze. On it spread, and the lingering storm came not to extinguish it. Gradually the crackling and roaring of the fire became terrific. Limbs fell with a crash, trees tottered and were thrown prostrate; the noise of falling timber was echoed from rocks and cliffs; and here, there, everywhere, rolling clouds of smoke were starred with burning cinders. On it swept for miles and scores of miles, from day to day, until more timber was destroyed than has been used by the people of Colorado for the last ten years.

Wildfires are equally spectacular today as they were in previous centuries. Traditional and social media inundate us with images of wildfires, especially when they burn homes, injure or kill people, and efforts to extinguish them. Vicariously watching heavy air tankers dump slurry in front of the flames or fire engines race down smoky roads from the safety of a mobile device or living room obscures both the power and danger of fire.

A Traumatizing Weapon

Because of its ability to rob people of possessions, waste resources, injure physically and psychologically, and kill, fire is a devastating weapon. Sun-Tzu, a Chinese military advisor for King Ho-Lu over 2,500 years ago and author of what is known today as *The Art of War*, outlined five ways in which fire can be used as an offensive weapon:

The first is to burn soldiers in their camp; the second is to burn stores; the third is to burn baggage-trains; the fourth is to burn arsenals and magazines; the fifth is to hurl dropping fire amongst the enemy.

He also identified rules regarding when fire could be used, which suggested a keen understanding of combustion. The proper season for attacking with fire is "when the weather is very dry" and days in which winds were expected to rise: when the moon is in the constellations of the Sieve, the Wall, the Wing or the Cross-Bar.

In *The Bible's* Book of Judges, Samson used fire to exact revenge against his father-in-law for giving his wife to another man:

> *And Samson went and caught three hundred foxes, and took firebrands, and turned tail to tail, and put a firebrand in the midst between two tails. And when he had set the brands on fire, he let them go into the standing corn of the Philistines, and burnt up both the shocks, and also the standing corn, with the vineyards and olives.*

The community retaliated with fire, but instead of attacking Samson, they burned the wife and her father. While many biblical references to fire represent divine power, in this case the locals were using a brutal tool to hurt one another.

This strategy, in which one army burns resources to prevent opponents from using them, has been used for centuries by the likes of Union General William T. Sherman in the American Civil War, Lord Herbert Kitchener in the South African Boer War, Joseph Stalin's retreating World War II army, and Russian armies retreating from Napoleon's advances. It's known today as the scorched-earth policy.

Greek fire is another form of weaponized fire that was used throughout the Mediterranean region between the 7th and 13th centuries. Its composition remains a mystery, but it could have contained resin, sulfur, saltpeter, oil or other flammable gels and liquids. It was usually thrown or sprayed onto enemy ships in naval battles, but versions of it also found their way into land battles.

Geoffrey de Vinsauf recorded the use of Greek fire during a crusade led by Richard I that started in 1187. In a sea battle against residents of Tyre, the crusaders aligned their boats so that their oars would entangle with the oars of the locals and create stable platforms for hand-to-hand combat:

> *...the oars become entangled and they fight hand to hand, having grappled each other's ships together; and they fire the decks with burning oil, which is vulgarly called Greek fire. That*

kind of fire with a detestable stench and livid flames consumes both flint and steel; it cannot be extinguished by water, but is subdued by the sprinkling of sand, and put out by pouring vinegar on it.

The Turks and other armies also used Greek fire against the crusaders, burning their wooden attack towers while attacking to defend their lands from the European invaders. Similarly, fire-pots were used during the 1415 siege of Harfleur, France. The city's defenders dumped containers of burning oils and fat against attacking Englishmen.

Napalm is among the most notorious forms of weaponized fire in recent memory. Naphthenic palmitic acid was invented at Harvard University in 1942, allegedly for use against buildings, but soldiers also targeted vegetation and people with the sticky, burning gel shot from handheld and vehicle-mounted flamethrowers as well as dropped in bombs from aircraft. Because napalm burned less rapidly than gasoline and other fuels, it tripled the range of flamethrowers and greatly increased the amount of burning fuel that reached its target. Since its creation it has seen action in most wars around the globe, incinerating multiple Japanese cities (such as Tokyo where over 87,000 died on March 9, 1945), defoliating jungles in Southeast Asia, and traumatizing civilians and soldiers alike in Europe, Asia and Africa.

Chemists manufacture napalm in factories, but fire itself is cheap and accessible to anyone who understands the fire triangle. As a result, it is a popular weapon among terrorists. Many recent terrorist attacks have involved fire, such as attacks on the US Embassy in Islamabad, Pakistan (1979), the US Embassy in Belgrade, Serbia (2008), the Taj Mahal Palace Hotel in Mumbai, India (2008), and the US Embassy in Benghazi, Libya (2012). Gunfire didn't kill US Ambassador Christopher Stevens and Communications Officer Sean Smith in that attack; they died of smoke inhalation. Closer to home, domestic terrorists burned the Two Elks Lodge on Vail Mountain in Colorado in October 1998.

Lasting Scars

One advantage to using fire as a weapon is that its delivery can be impersonal. It can be delivered from aircraft. It can be ignited from upwind or downslope of the target. The person wielding fire doesn't have to see any victims.

In 1944, Japan initiated its Fugo Campaign against the United States, launching large balloons that carried incendiary bombs into the jet stream. Its military leaders hoped to ignite forest fires that would "create havoc, dampen American morale and disrupt the U.S. war effort." Of the 9,000 balloons launched, only a few hundred reached North America, but none ignited fires. They did travel as far east as Kalispell, Montana; Detroit, Michigan; and Omaha, Nebraska.

Arsonists also wield fire as a weapon to exert control over others, to inflict pain, to relieve their own suffering, or to produce excitement at others' expense. While attending the 2015 Fire & Life Safety Educators Conference of the Rockies, I listened to George Keller share a story about a serial arsonist who tormented Washington State in 1992 and 1993. That arsonist was his then 27-year-old son Paul.

Paul Keller set at least 76 fires, damaging homes, businesses, churches and an assisted living community and killing at least three people. A clinical psychologist who examined Keller was quoted in the *Seattle Times* as saying fire "endowed the weak child, Paul Keller, with power." He felt excitement watching fires burn buildings and watching firefighters race through their districts to battle them. At one point, George Keller told us, Paul waited for firefighters to begin extinguishing one fire before he ignited another building across the street.

Paul Keller destroyed more than buildings and property. He destroyed memories, stability, security, relationships and trust. "You can rebuild the walls," George Keller said at the conference, "but the scars last forever."

That's what fire does. Fire sears, incinerates, traumatizes and kills. Yet the first green sprouts of new growth following a fire—an organic phoenix—reveal the complement to fire's destructive power: it also creates. ᐧ

5

Fire's Positive Side: A Tool for Creation

It was the private soldier who taught me... how to light a fire in the woods with wet twigs in a pelting rain and a fretful wind with your last match.

Private Robert J. Burdette, 47th Illinois Infantry Regiment, 1862

They dig a hole in the ground sufficiently large to contain the fire, and then they put sticks in the ground, on the margin of the hole, so as to bring the tops of them together, and fasten them; the meat is placed on the sticks above, and the fire from the hole beneath cooks it quite well.

Sarah Ann Horn, 1839

*W*hen thinking about this complex topic, I realized that I've been around fire and its transformative powers all of my life. My dad, maternal granddad and other relatives spent part of their lives finding and mining molybdenum. It has multiple industrial uses, but the one that made the most sense was its value in producing steel; fire blends molybdenum with iron and other metals to create stronger and lighter steel products.

Pottery was another route through which I learned about fire. My mom created pottery and she often used our own fireplace for her effort. Once she ran afoul of local authorities when, as the volunteer elementary school art teacher, she built a San Ildefonzo Pueblo-style kiln in a lot near my school to heat the clay pots we made in her art class. Again, fire was at work changing moist clay into hard pottery.

Maria Martinez made that style of pottery famous. Born in 1881, she emulated the style of pottery that her aunt and countless ancestors had practiced. She rolled coils of clay and stacked them to create a tall cylinder. She shaped the cylinder into different vessels with more gradual contours and used smooth river rocks to polish the clay's surface. After the pot dried, it was baked in a wood fire. Maria and her husband Julian learned to smother the fire with manure to produce blackened hues.

Clay is essentially decomposed granite, which is a combination of quartz, feldspar, mica and minerals such as alumina and silica. The way in which granite decomposes over millions of years determines the quality and characteristics of the resulting clay; each substance can vitrify (melt) when exposed to varying amounts of heat. After a potter—or elementary art student—shapes a lump of clay into a desirable form, they place those forms into a kiln, a container in which heat can be generated in sufficient quantities to make minerals and quartz vitrify. Vitrification typically occurs between 2,200 and 2,700 degrees Fahrenheit. If the temperature is too high or applied too long, clay can melt into a puddle. On the other hand, incomplete vitrification creates pores in the fired clay, which is not ideal when creating vessels for liquids.

Even after my mom bought her own kiln for the garage, she experimented with tin cans in our fireplace to convert smaller clay forms into hardened pots in a fashion similar to Maria Martinez.

Just as I learned a lesson about fire and its usefulness from my mother, the Choctaw people credit another female for sharing the lesson of pottery as well as other benefits and warnings regarding fire. According to the following story, Grandmother Spider stole fire from the people of the East and shared it with the Choctaws:

> The Choctaw People say that when the People first came up out of the ground, People were encased in cocoons, their eyes closed, their limbs folded tightly to their bodies.
>
> And this was true of all People, the Bird People, the Animal People, the Insect People, and the Human People. The Great Spirit took pity on them and sent down someone to unfold their limbs, dry them off, and open their eyes.
>
> But the opened eyes saw nothing, because the world was

dark; no sun, no moon, not even any stars. All the People moved around by touch, and if they found something that didn't eat them first, they ate it raw, for they had no fire to cook it.

All the People met in a great powwow, with the Animal and Bird People taking the lead, and the Human People hanging back. The Animal and Bird People decided that life was not good, but cold and miserable. A solution must be found! Someone spoke from the dark, "I have heard that the people in the East have fire."

This caused a stir of wonder, "What could fire be?" There was a general discussion, and it was decided that if, as rumor had it, fire was warm and gave light, they should have it too. Another voice said, "But the people of the East are too greedy to share with us..."

Then a small voice said, "We will take it, if Grandmother Spider will help." The timid humans, whom none of the animals or birds thought much of, were volunteering!

So Grandmother Spider taught the Human People how to feed the fire sticks and wood to keep it from dying, how to keep the fire safe in a circle of stone so it couldn't escape and hurt them or their homes. While she was at it, she taught the humans about pottery made of clay and fire, and about weaving and spinning, at which Grandmother Spider was an expert.

Fire from Down Under

As several ancient stories reveal, fire can be beneficial. It renews ecosystems, cooks food, and warms us against the cold. Australia's Aborigines have a rich canon of stories that teach the importance of fire. The first of the following four stories wasn't attributed to a specific area, but the second excerpt is from the northern coast of what is now New South Wales and the third is from what is now Victoria. The last story is also from New South Wales.

There was a time when the Australian bush was different from what it is to-day. Trees were bigger and their wood softer. There were more and bigger and brighter flowers. And the land—especially the mountains—was far more densely clothed with verdure. But some change came, and it was not good for the land. Seeds failed to germinate, and where fertile tracts had been now desert appeared...

...Again the maiden thought of beseeching the spirit. She went back to the old ground all alone and she found the clay. She painted herself and awaited results. She heard the spirit and she talked with it. Then she noticed that just before her a little smoke wreath curled up into the air. Then a flame burst, and in a very little while a fierce bush fire was raging.

The girl was satisfied that a fire was what was needed and she sent word to the river to say that all would soon be well with the world. That the seeds would germinate and new plants would grow up and flower and all would be gay as before.

Since that time bush fires do not need any mystic markings nor special communings by special people. Limbs of trees rub themselves hot on dry days and make flame. The hot sun shining on the mica in the rocks set fire to the tiny mosses that are dried there. And so without human agency the fires come that are necessary to make our Australian seeds burst into the life of a new and growing plant.

* * *

Long ago, before even the Dreamtime, there was a tribe of people who did not live on the earth. They lived in the sky world and their camp was near the two brightest stars so that they could light their fire-sticks from them. They were the only people anywhere who had the use of fire. The people on earth had to manage without it...

Two brothers, named Kanbi and Jitabidi, brought their fire-sticks down to earth with them and left them smoldering while they went off hunting. Hunting possums turned out to be a lot more difficult than they had expected and the time drew out and the land was very quiet. The fire-sticks became bored and began to play 'chase'. They were very clever at this game, running from place to place, and everywhere they touched the dry grass it caught alight. Gradually all the little fires grew together into one big fire and the smoke could be seen from a long way off. As soon as the sky brothers saw the smoke they left the hunting party and hurried back to put out the fires.

The Aboriginal people who lived in the area had also seen the smoke and had come to see what was going on. They had never seen fire before and at first they were very afraid. It did not take them long, however, to realize that this strange phenomenon could be extremely useful to them, providing them with light and warmth. They also noticed that some possums the sky brothers had caught had been cooked by the fire and smelled wonderful and savory. They realized that they too could make their food more tender and tasty with fire...

Kanbi and Jitabidi quickly gathered up their playful firesticks and returned to their campsite in the sky. They were terribly afraid the earth people would inflict some punishment on them for having caused such a disturbance. But the earth people were in awe of their sky visitors and rather than being angry about the burnt grass were excited and grateful for the wonderful gift of fire.

* * *

The bandicoot was once the sole owner of fire, and cherishing his fire-brand, which he carried with him wherever he went, he obstinately refused to share the flame with anyone else. Accordingly the other animals held a council and determined to get fire either by force or by stratagem, deputing the hawk and the pigeon to carry out their purpose. The latter, waiting for a favorable moment when he thought to find it unguarded, made a dash for it; but the bandicoot saw him in time, and seizing the brand, he hurled it toward the river to quench it. The sharp eyes of the hawk saw it falling, and swooping down, with his wing he knocked it into the long dry grass, which was thus set alight so that the flames spread far and wide, and all people were able to procure fire.

Interestingly, the bandicoot in this story is replaced by two greedy women in the next, which might say something about the storyteller and his opinion of women.

> *...Fire was originally owned by two women (Kangaroo-Rat and Bronze-Winged Pigeon) who kept it concealed in a nut-shell. For a long time the other animals could not discover how these women were able to cook their food; but at last they set spies to watch them and so learned the secret, whereupon, re-solving to secure fire by a ruse, they arranged a dance and invited the two women to be present. One after another the different animals danced in ludicrous positions in an attempt to make the women laugh; and at length one performer suc-ceeded so that the women, convulsed with merriment, rolled upon the ground. This was just what the conspirators had been waiting for, and rushing up, they seized the bag in which was the nut that contained the fire. Opening this and scattering the flame about, they set the grass alight, and in this way fire was caught in the trees, whence ever since it can be procured from their wood by means of friction.*

According to the Karoks of what is now northwestern California, fire was a prized possession of spirits known as skookums. It was so prized that they refused to share it, but Coyote had other plans:

> *In the beginning, the animal people had no fire. The only fire anywhere was on the top of a high, snow-covered mountain, where it was guarded by the skookums. The skookums were afraid that if the animal people had any fire they might become very powerful, as powerful as the skookums. So the skookums would not give any of the fire away to anyone.*
> *Because the animal people had no fire, they were always shivering, and they had to eat their food raw. When Coyote came along he found them cold and miserable.*
> *"Coyote," they begged, "you must bring us fire from the mountain or we will one day die of all this cold."*
> *"I will see what I can do for you," promised Coyote...*
> *[After stealing fire from the skookums] Coyote then called*

the animals together and they all gathered around Wood. Coyote, who was very wise, knew how to get the fire out of Wood. He showed the animals how to rub two dry sticks together until sparks came. Then he showed them how to collect dry moss and make chips of wood to add to the sparks to make a little fire. Then he showed them how to add small twigs and pine needles to make a bigger fire.

From then on, the people knew how to get the fire out of Wood. They cooked their meat, their houses were warm, and they were never cold again.

This story is similar to one told by the Shastas who live in a similar region, but they replaced skookums with a sisterhood of Fire Beings, again suggesting that gender was as important then as it is now.

Modern humans take cooking for granted, but according to several stories from around the world, eating raw meat or relying on a vegetarian diet of grasses was the norm for generations until the people—such as the Motus of New Guinea and the Dyaks of Borneo—and their non-human allies found, stole or otherwise acquired fire and its benefits.

The ancestors of the present people had no fire, and ate their food raw or cooked it in the sun until one day they perceived smoke rising out at sea. A dog, a snake, a bandicoot, a bird, and a kangaroo all saw this smoke and asked, "Who will go to get fire?" First the snake said that he would make the attempt, but the sea was too rough, and he was compelled to come back. Then the bandicoot went, but he, too, had to return. One after another, all tried but the dog, and all were unsuccessful. Then the dog started and swam and swam until he reached the island whence the smoke rose. There he saw women cooking with fire, and seizing a blazing brand, he ran to the shore and swam safely back with it to the mainland, where he gave it to all the people.

* * *

One day when the man and the dog were in the jungle to-gether, and got drenched by rain, the man noticed that the dog warmed himself by rubbing against a huge creeper (called the Aka Rawa), whereupon the man took a stick and rubbed it rapidly against the Aka Rawa, and to his surprise obtained fire. Later some food was accidentally dropped near the fire, and the man, finding it thus rendered more agreeable to the taste, discovered the art of cooking.

Wildfire Is a Tool

In light of the large wildfires burning across the West over the last few years, it is noteworthy that ancient wildfires were seen as positive forces in their ecosystems: "The girl was satisfied that a fire was what was needed and she sent word to the river to say that all would soon be well with the world." For the early cultures of Australia, as well as native cultures across the world, wildfires were not viewed as we see them today: threatening and negative.

In fact, wildfire was a tool, but it wasn't the same tool for everyone. Diverse Native cultures understood and used fire according to their own economic and ecological needs. Farmers used it differently than hunter gatherers. People who lived in mountains had different needs than those who lived on the Plains. As Native American groups moved from eastern ecosystems into midwestern and western ones, they had to adapt their use of fire just as they adapted their cultures to their new ecological surroundings. Technological changes, such as adopting horse culture, also changed fire use.

Fast-forward to the late 19th century and human relationships with fire had changed again. Explorer John Wesley Powell understood the benefits of wildfire when, in 1890, he wrote about the nation's non-irrigable lands for *The Century Magazine*:

> *It is thus that, under conditions of civilization, the great forests of the arid lands are being swept from the mountains and plateaus. Before the white man came the natives systematically burned over the forest lands with each recurrent year as one of their great hunting economies. By this process little destruction of timber was accomplished; but, protected by civilized men, forests are rapidly disappearing. The needles, cones, and brush, together with the leaves of grass and shrubs below, accumulate when not burned annually. New deposits are made from year to year, until the ground is covered with a thick mantle of inflammable material. Then a spark is dropped, a fire is accidentally or purposely kindled, and the flames have abundant food.*
>
> *There is a practical method by which the forests can be preserved. All of the forest areas that are not dense have some value for pasturage purposes. Grasses grow well in the open grounds, and to some extent among the trees. If herds and flocks crop these grasses, and trample the leaves and cones into the ground, and make many trails through the woods, they destroy the conditions most favorable to the spread of fire. But if the pasturage is crowded, the young growth is destroyed and the forests are not properly replenished by a new generation of trees. The wooded grounds that are too dense for pasturage should be annually burned over at a time when the inflammable materials are not too dry, so that there may be no danger of great conflagration.*

However, Powell's grasp of fire science and its benefits to ecosystems wasn't universal.

As many Euro-American settlers emigrated from forested countryside into the Great Plains, they found themselves ecologically dislocated. They missed their trees and misunderstood the role fire had in the ecosystem. In conjunction with massive herds of roaming ungulates, wildfires had created the fertile tall-grass and short-grass prairies of the Midwest. Excluding fire from the ecosystem changed the prairie.

One Oklahoma settler recalled the changes on his family's land following their arrival in 1908:

> *There used to be big fires. That was one thing that fooled my dad. He thought he was getting a prairie place more or less. After he started pasturing it...why the trees commenced to growing [and] in eight to ten years it all turned to brush.*

Both brush and trees invaded prairies when wildfires stopped burning as regularly. Scientist Charles Bessey interviewed several farmers in eastern Nebraska about trees extending their habitat into formerly treeless prairies:

> *Another says that in 1872 very few of the 'draws' (i.e. ravines) had any trees in them, but now where fire is kept out all are filled with timber. He says that on his farm which was originally swept with prairie fires, I had a draw where water was half the year, in which in 1883 there were no trees of any kind, while now there are willow, cottonwood, box elder and elms... the timber belt along the Nemaha River has widened from a hundred feet to half a mile.*

Those farmers recognized that fire exclusion was leading to tree growth, but they didn't seem to mind because many of them believed that trees produced wet soils: "...cause springs to break forth, supply wells with never-failing water, and make the creeks flow all the year round."

In fact, they waged their own war against fire because it endangered their families, homes, equipment and crops. They also cursed it as a symbol of native, non-Christian, nomadic North America: "or shall we be savages and apply the burning torch to every blade of grass?" Because fighting fire was a cultural war as much as an economic one, there was little to be learned from the enemy and the people who wielded it. Siting a community along a stream or along bluffs may have had as much to do with utilizing natural firebreaks as with access to resources or max-imizing line of sight to see welcome and unwelcome visitors.

Countless historic observations and current tests have demon-strated that fire is essential in many ecosystems. Land managers cite such evidence when they allow some wildfires to burn as project fires, and ignite their own wildfires—known as prescribed fires—to vitalize and modify ecosystems.

Prescribed fire is used today, but under tight regulations. These fires are managed to mimic the effects of low intensity surface fires: recycling nutrients (primarily carbon and trace minerals) from veg-etation into the soil, shaping populations of targeted plants, creating edge habitat, and protecting resources, infrastructure and homes from wildfires. They also provide live-fire training to firefighters and living laboratories for scientists.

Despite its benefits, prescribed fire isn't used as much as it could be. Prescribed fires that escape and make headlines, such as New Mexico's 2000 Cerro Grande and Colorado's 2012 Lower North Fork tragedies, are part of the problem. The Coalition of Prescribed Fire Councils iden-tified several obstacles in a 2012 report: capacity (personnel trained to conduct prescribed fires), weather, air quality concerns, resources, public perception, liability, permitting, and population growth into fire-prone ecosystems.

Fire is both destructive and creative, often simultaneously. Using it and understanding it require rules, both for the individual and for communities. ☞

6

Rules of Fire, Rites of Fire

...The fireflies pursued Fox to his burrow and informed him that, as punishment for having stolen fire from them and spreading it over the land, he should never be permitted to use it himself.

Jicarilla Apache legend

The work of fire prevention may impress you as a fad or a joke at the outstart, but your matured view will be that it is both interesting and serious... This new responsibility calls for intelligence, sound judgment, broad views and amenable to reason. Narrow mindedness and bombast have absolutely no place in this work and every new and practical idea bearing on the end ultimately to be achieved should be promptly grasped and utilized.

Philadelphia Fire Department, 1913

*R*ites are solemn responsibilities, often sacred in nature. A rite of passage is a significant, ceremonial event in a person's life that is common among a family, society or culture: graduating from high school, catching a first fish, climbing a 14,000-foot peak, bat mitzvah, a first kiss, sharing a campfire, or attacking a structure fire. For some cultures learning how to acquire and harness fire was a rite of passage from childhood to adulthood, or from achieving as an individual to contributing to the community. Such rites are one theme of this chapter.

Rules about fire use are the other theme here. Fire existed without humans through lightning, magma and objects landing on the planet after grinding through the atmosphere, yet it also can be created by humans when we combine a proper mixture of heat, fuel and oxygen.

Fire can be dangerous if a person acquires it without simultaneously acquiring cultural instructions, such as how to avoid getting burned, how to limit the heat, and how to protect it.

I'm reminded of coursework in graduate school that interpreted interwoven histories of humans, bison and horses. Horses had evolved in North America, but died out as part of the megafauna extinctions after humans left Asia, crossed Beringia and entered their New World approximately 13,000 years ago. Megafauna, including ancestral horses, were unable to adapt quickly enough to the perfect storm of new two-legged hunters and changing climatic conditions. They died out. As a result, humans developed pedestrian cultures over the next dozen millennia.

When European traders, missionaries, conquistadors and trappers brought horses back to North America starting in the late 1400s, indigenous peoples didn't know how to utilize the new creatures until they learned the accompanying culture. In essence, the horse was technology. New technology will sit on the shelf or wreak havoc if it doesn't come with the equivalent of an instruction manual.

For nomadic Plains Indians, the transition from pedestrian to equestrian cultures occurred in less than two generations, but it did not occur in a single afternoon or overnight. They had to learn basic horse maintenance such as first aid, feeding and corralling. They learned riding skills and packing skills. They learned how to hurl spears, shoot arrows and later shoot guns from the back of a sprinting horse. They taught their horses how to be calm in battle and while running along stampeding herds of bison. Adopting horse technology was a significant investment for those first generations, yet I'd bet later generations quickly took horses for granted.

The same is true with fire. It is a form of technology. It can make food edible, refine valuable metals, transform mud into pottery, heat our hearths, and replace icky stenches with pleasant scents. When it is taken for granted, it destroys, injures and kills as much as it creates and recycles.

Researchers recently found evidence of Neanderthals using fire habitually at least 400,000 years ago, based on remains of charcoal, heated stone artifacts, burned bones, heated sediments and hearths at potential fireplace sites. They also found evidence of our ancient ancestors burning bark peels to create pitch needed to fit wooden shafts on stone tools.

According to the people of Sulawesi, an island formerly known as Celebes, fire was given by the deity to the first men, but they allowed it to go out. Since they had not learned the culture to go with the fire and thus how to ignite a new fire, they sent a man named Tamboeja to the deities in the sky to get flame. The inhabitants of the sky-world told him that they would give him fire, but that he must cover his eyes with his hands so that he would not see how it was made. They were unaware that Tamboeja conveniently had eyes in his arm pits that enabled him to watch their actions and see how they made fire with flint. Upon returning to his community he shared the fire and the secret of the flint—an oral quick reference guide—with his people.

Humans have acquired fire a number of ways, from outright theft to receiving it as a gift. In most stories, such as this one from the Nigeria, the gifting is documented, but without any overt rules for tending fire. The Yorubas acquired fire from Arámfè, the God of Thunder and Father of all Gods.

> *...Odudúwa sent*
> *Ífa, the Messenger, to his old sire*
> *To crave the Sun and the warm flame that lit*
> *The torch of Heaven's Evening and the dance. . .*
> *A deep compassion moved thundrous Arámfè,*
> *The Father of the Gods, and he sent down*
> *The vulture with red fire upon his head*
> *For men; and, by the Gods' command, the bird*
> *Still wears no plumage where those embers burned him—*
> *A mark of honour for remembrance.*

The lack of a user's guide may be why we're still having issues with fire today; we have to keep learning old lessons.

Modern Americans wield fire, from outdoor grills and birthday candles to cigarettes and internal combustion engines, with little thought to its physical power and even less thought to its cultural history. We take it for granted without considering its origins and how we came to harness it or at least have the capacity to harness it.

Children in particular lack an awareness of fire's power and history. They see adults wield fire casually and attempt to emulate their role models. They see adults put cigarettes—burning paper, leaves and chemicals—within inches of their faces. They see adults start campfires in xeric forests. They see shortened recordings of live-fire escapades on the Internet. They don't see the chronic injuries and illnesses, experiences in the burn units of hospitals, or the preparations to make those fire experiences as safe as possible. They don't see destroyed property or fathom the emotional losses associated with that property.

Many adults lack such an understanding as well. They burn yard waste during hot, dry and windy periods and build combustible homes in wildfire-prone ecosystems. They burn ditches during windy conditions. They use fireworks during burn bans and launch model rockets over fields of dry grass. They leave matches and lighters within arm's reach of children at home, in playgrounds and in convenience stores. Ignorance and carelessness certainly walk arm in arm in these situations. Much like an axe or hammer, fire is a helpful tool that can be dangerous in untrained (or sinister) hands.

Often our ancestors learned these rules from stories of their ancestors making good and bad decisions. In the following Lnu'k story from the northern part of Maine, Wolverine was imprudent and abused his new access to fire. He nearly died as a result:

One day Wolverine visited his older brother Bear, who was very glad to see him, and at once put the pot on the fire to cook him something. After the food was cooked and they had eaten it, Bear said to his younger brother Wolverine, "How would you make a fire if you did not have any flint and steel?" Wolverine acknowledged that he would be helpless without flint and steel. "Now I will teach you," said Bear, "how to make a fire, when you do not have any flint and steel." Having said this, Bear went out and got some maple bark, which he put in a little pile, and then jumped over it. As soon as he jumped over it, it burst into a flame. Then he said to his younger brother, "Now I give you power to make a fire."

Wolverine was very happy and was in a hurry to get away and try his power. As soon as he got out of the house, he started to run. He continued running until he got to a place where he could no longer see Bear. Then he collected some maple bark and made a little pile of it and jumped over it. When it broke into a blaze, he was very much pleased. He took out his flint and steel and threw them away, saying, "These are no longer of any use."

Wolverine had no use for the fire he made; he only made it to try his power. So he went on, but he had hardly gotten out of sight of his first fire, when he decided to make a new fire. After that he made fires more frequently until at last he made them every ten steps; but finally his power gave out, for he had used it all up. When he next collected a pile of maple bark and jumped over it, it did not burst into flame. By that time it had grown dark and was very cold, and he was indeed in need of a fire. Then truly he jumped, but no success crowned his efforts. He had thrown away his flint and steel and was very much frightened, for it was very cold. He kept on jumping, but it grew so cold that he froze to death while he was jumping. He lay there until spring, when he thawed out…

Other ancient stories also capture the rules that could be part of a fire instruction manual. The Alabamas of southeastern North America added burn prevention messaging to their stories:

Once a Fire almost burned out and was making the hissing sound usually heard. When a man asked it what it wanted, the Fire said, "Something to eat."
"What do you want to eat?"
"I want to eat wood."
So the man picked up some dead wood and piled a quantity of it on the Fire. The Fire grew bigger and bigger, and the man kept piling on more and more wood, until the Fire cried to the man to hold all of the animals back so that they would not be burned.

* * *

Bears formerly owned the Fire and they always took it about with them. One time they set it on the ground and went farther on, eating acorns. The Fire nearly went out and called aloud.

"Feed me," it said.

Then some human beings saw it. They found a stick north of the fire and placed it over the flames. They found another stick west and placed it on top. They found a stick south of the fire and finally east of the fire, and placed them on top in order. The Fire blazed up.

When the bears came to get their Fire, it said, "I don't know you anymore." They did not get it back and so it belongs to human beings.

I tell a version of this story to 4th graders and ask them what rules the humans need to follow in order to keep themselves safe. Most students emphasize the need to limit fuel and stay close, but not too close, to fires when they are burning. In other words, when you're cooking, stay in the kitchen, and don't abandon campfires until the fire is out completely. When shared appropriately, these old rules resonate in new audiences.

A Constant Threat to Pioneers

Most of the men and women who migrated west in the latter half of the 19th century were poorly equipped in terms of knowledge and equipment for such transcontinental expeditions. Captain Randolph Marcy wrote *The Prairie Traveler* to address many of those issues. He had earned his credibility and his post as Inspector General of the Department of Utah after fighting in the Mexican War, escorting westbound emigrants, locating military posts, exploring the wilderness, and accompanying the expedition against Mormons in Utah. He also led his men safely from Utah to New Mexico on a forced winter march through the Rocky Mountains without adequate provisions.

He recognized wildfire as a threat and shared tips on combating it in his handbook:

Inexperienced travelers are very liable, in kindling fires at their camp, to ignite the grass around them. Great caution

*should be taken to guard against the occurrence of such acci-
dents, as they might prove exceedingly disastrous. We were
very near having our entire train of wagons and supplies destroyed, upon one occasion, by the carelessness of one of our
party in setting fire to the grass, and it was only by the most
strenuous and well-timed efforts of two hundred men in setting counter fires, and burning around the train, that it was
saved. When the grass is dry it will take fire like powder, and
if thick and tall, with a brisk wind, the flames run like a race-
horse, sweeping everything before them. A lighted match, or
the ashes from a segar or pipe, thrown carelessly into the dry
grass, sometimes sets it on fire; but the greatest danger lies in
kindling camp-fires.*

*To prevent accidents of this kind, before kindling the fire a
space should be cleared away sufficient to embrace the limits of
the flame, and all combustibles removed therefrom, and while
the fire is being made men should be stationed around with
blankets ready to put it out if it takes the grass.*

*When a fire is approaching, and escape from its track is
impossible, it may be repelled in the following manner: the
train and animals are parked compactly together; then several
men, provided with blankets, set fire to the grass on the lee side,
burning it away gradually from the train, and extinguishing it
on the side next the train. This can easily be done, and the fire
controlled with the blankets, or with dry sand thrown upon it,
until an area large enough to give room for the train has been
burned clear. Now the train moves on to this ground of safety,
and the fire passes by harmless.*

Marcy recognized the danger of fire as well as its limits. He may not
have used the term Fire Triangle, but he knew that fuel was essential.
Dry fuels propagated fires, especially in windy conditions, and eliminating fuel, such as by setting "counter fires" or clearing a space of
combustibles around a campfire, stopped fires from growing.

Wildfire was a persistent threat to pioneers in forest and prairie
ecosystems. Even as a teenager, Luna E. Warner recorded her concerns
about fire in her diary entries in October 1871:

October 12: I went down across the river. Saw a turkey and a squirrel. Pa and I went up the river looking for game. Ever so many geese went over. Some stopped in the river. Mr. Root went to Cawker with Uncle Eli. He heard that a great part of Chicago is burnt and the fire is still burning. Alf shot a turkey down by the river. . .

October 20: Louie went over to see a herd of 1,300 Texas cattle that went up the river on the other side. There are fires burning in all directions every day.

October 21: After noon I went over to Uncle Eli's and stayed to supper. Then came home and got the boots for Velma to wear across the river. She braided her hair when she got here. Pa tried to burn a strip to keep off fire but it was so still it would not burn. We had turkey for dinner. The folks played high, low, jack all evening.

October 28: I saw 11 turkeys coming towards the house. Pa went out and shot one dead-a young gobbler. Alf, Arabella and I started for Cawker. There was a prairie fire opposite Youngs'. We saw Miss Lines. We had dinner with Gena. Got home before dark.

October 30: What a relief to have Root gone! Mama and I dug potatoes. There was a prairie fire that came down to Uncle George's claim. Uncle Eli and Venelia came along. Venelia stayed with me. They brought word that Uncle Howard is in Solomon. Alf will go after him in the morning.

November 1: After noon we saw a fire coming in the bottom beyond Ray's. We all went out and set fires all around the ploughing. Before we got the road burnt, the fire came sweeping down from Beal's, swept across the road into the bottom. Then we burnt side of the path to the river and kept burning until night and made out to save the premises. We were tired...

November 2: Arabella and I went over to Uncle Eli's. Had a splendid vegetable dinner. We came home by the ford past Arabella's house. It is burned clear to the river. Several trees burned.

Clearly rules are important, but sometimes rules must be broken such as when Bobok, Coyote or Grandmother Spider stole fire in cultures that don't approve of stealing. In the following case from Samoa, Ti'iti'i broke the rule that fire had to be used for destruction when he used fire to cook food. As a trickster figure, he creates good from bad and manipulates the listeners' and readers' expectations about expected consequences of breaking divine rules.

> *The home of Mafuie, the earthquake god, was in the land of perpetual fire. Ti'iti'i's father Talanga was also a resident of the under-world and a great friend of the earthquake god...*
>
> *In a little while the boy saw smoke and asked what it was. The father told him that it was the smoke from the fire of Mafuie, and explained what fire would do.*
>
> *The boy was determined to get some fire. He went to the place from which the smoke arose and there found the god, and asked him for fire. Mafuie gave him fire to carry to his father. The boy quickly had an oven prepared and the fire placed in it to cook some of the taro they had been cultivating. Just as everything was ready an earthquake god came up and blew the fire out and scattered the stones of the oven.*
>
> *Then Ti'iti'i was angry and began to talk to Mafuie. The god attacked the boy, intending to punish him severely for daring to rebel against the destruction of the fire.*
>
> *A severe battle ensued. At last Ti'iti'i seized one of the arms of Mafuie and broke it off. He caught the other arm and began to twist and bend it.*
>
> *Mafuie begged the boy to spare him...*
>
> *Ti'iti'i listened to the plea and demanded a reward if he should spare the left arm. Mafuie offered Ti'iti'i one hundred wives. The boy did not want them.*
>
> *Then the god offered to teach him the secret of finding fire to take to the upper-world. The boy agreed to accept the fire secret, and thus learned that the gods in making the earth had concealed fire in various trees for men to discover in their own good time, and that this fire could be brought out by rubbing pieces of wood together.*

Inclusion and education about rules for using fire are the best ways to reduce injuries and property loss because they solve questions borne of curiosity and reduce experimentation.

For 50 years, the Independence Day fireworks display was part of the identity of the fire department in Idaho Springs, Colorado. While residents and thousands of visitors embraced the show as a summer tradition, I learned it was a rite of passage within the Idaho Springs Fire Department (ISFD).

Chief Don Krueger first asked me to stay on "The Hill" for the ignitions process in 1999, only my second summer on the fire department. I had helped collect money from the canisters that year and helped hang the Niagara Falls display in the afternoon, in addition to responding to a large percentage of the emergencies that spring and the previous year. The invitation to join the crew on The Hill was a treasured mark of respect in the department. Many firefighters must remain in the valley, collecting donations and patrolling for fires and illegal fireworks. The opportunity to work on The Hill was almost too much—I actually considered refusing the offer out of a sudden burst of shyness—but friends convinced me to "get up there." I went. Thank goodness, I went.

Once I found a ride onto The Hill, I was introduced to "Uncle Bob." Bob Albers, a former ISFD volunteer firefighter himself, coordinated the show, deciding when to launch shells from the respective tubes, when to light the set pieces or boxes, and when to ignite the waterfall display. Chief Krueger introduced him to me as Uncle Bob, and I've called him by that name ever since. Six middle-aged firefighters and long-time members of the ISFD completed the crew. My job was helping two of them load their three- and four-inch mortars. At the appropriate time, all of us donned our bunker gear and slipped foam plugs into our ears to muffle the upcoming assault on our eardrums.

I was so busy that I hardly caught any of the show, racing between the mortar tubes and the metal trash cans holding the remaining shells. It was tough work with lots of sweating and an occasional misfired shell exploding only a few inches above our heads instead of several hundred feet above the valley. I had to shield our shells from the drizzle of sparks and smoldering shrapnel falling from above, tucking them close to my body and remembering to replace the lids on the cans before racing back to the tubes. As chaotic as the rite was, there were rules to follow: wear your bunker gear, use eye and ear protection, protect the shells from sparks, listen to directions.

Those memories of fire's combination of beauty and potential violence influence my current fireworks messaging when kids and adults inquire about whether a device is legal in South Metro Fire Rescue's district. Fireworks appear safe largely because of the lack of personal feedback. Each time a firecracker is used safely, it gives an impression that every interaction will be safe. Even those relatively safe sparklers burn at 1,200 degrees Fahrenheit. As my family learned a long time ago, all fireworks are dangerous devices.

It was dusk and we had walked the half-block from our home to the elevated lawn of the parsonage. It was a perfect spot. When the big show started we need only look straight up and the whole night sky would glitter and zoom for hours (at least it seemed like that to me).

My dad had just lit up one of the best fireworks ever. First it smokes in lovely clouds of color that go whizzing around in circles and then it booms and explodes in an infuriatingly surprising ball of light.

This particular device had a mind of its own. I was in my mother's lap on a blanket in the grass and sitting cross legged as the smoke bomb started its dance. It whizzed in yellow and green. It spun and twirled and I was mesmerized. And then it started to bounce. The delicate, smoky ballet was coming to a fantastic finale as it twirled and bounced through the air.

And then it landed in my lap. Fuse, fire and all. I don't really remember what happened next except that there was an explosion, a very loud pop a few feet away from me, and I was crying and everyone on the lawn was asking me if I was okay. But I was okay. Someone tossed it back out to the street.

My earliest Fourth of July memory is the moment when I didn't get blown up.

Ignorance, Casualness, Distractions

Because fire is so dangerous, even firefighters—the men and women called upon to react to a fire to curb its damage and extinguish it—have rules of engagement. Wildland firefighters know the basic rules as the "10 & 18."

The original 10 Standard Firefighting Orders were developed in 1957 by a task force commissioned by the USDA-Forest Service Chief Richard E. McArdle. The task force reviewed the records of 16 tragedy fires that killed 79 firefighters from 1937 to 1956 including the Welcome Lake (1937), Mann Gulch (1949), Rattlesnake (1953) and Inaja (1956) incidents. They identified common factors among those incidents and created standard firefighting orders "to be committed to memory by all personnel with fire control responsibilities" to prevent future tragedies. Based on the military's "General Orders," the Standard Firefighting Orders were organized deliberately to be implemented systematically and applied to all fire situations to assure that firefighters paid attention to fundamentals of safety.

The verbiage of the orders have changed over time, but the substance has not.

Standard Firefighting Orders

1. Keep informed on fire weather conditions and forecasts.
2. Know what your fire is doing at all times.
3. Base all actions on current and expected behavior of the fire.
4. Identify escape routes and safety zones and make them known.
5. Post lookouts when there is possible danger.
6. Be alert. Keep calm. Think clearly. Act decisively.
7. Maintain prompt communications with your forces, your supervisor, and adjoining forces.
8. Give clear instructions and insure they are understood.
9. Maintain control of your forces at all times.
10. Fight fire aggressively, having provided for safety first.

Shortly after the Standard Firefighting Orders were incorporated into firefighter training, the "13 Situations That Shout Watch Out" were developed. They also resulted from killer fires.

Now numbering 18, these rules of engagement are more specific than the Standard Firefighting Orders and described situations that expand upon the original ten.

18 Watchout Situations

1. Fire not scouted and sized up.
2. In country not seen in daylight.
3. Safety zones and escape routes not identified.
4. Unfamiliar with weather and local factors influencing fire behavior.
5. Uninformed on strategy, tactics, and hazards.
6. Instructions and assignments not clear.
7. No communication link with crewmen/supervisors.
8. Constructing line without safe anchor point.
9. Building fireline downhill with fire below.
10. Attempting frontal assault on fire.
11. Unburned fuel between you and the fire.
12. Cannot see main fire, not in contact with anyone who can.
13. On a hillside where rolling material can ignite fuel below.
14. Weather is getting hotter and drier.
15. Wind increases and/or changes direction.

16. Getting frequent spot fires across line.
17. Terrain and fuels make escape to safety zones difficult.
18. Taking a nap near the fire line.

When firefighters follow the Standard Firefighting Orders and remain alert to the Watchout Situations, much of the risk of firefighting is reduced. It sounds easy, but firefighters get injured and killed on wildfire incidents every year. Nineteen members of the Granite Mountain Hotshot crew died in the Yarnell Hill Fire in June 2013; based on early comments from the investigation, some of the Watchout situations were factors.

Former Zig Zag Hotshot Supervisor Paul Gleason, a guru of the wildland fire realm, addressed the dangers of firefighting in a presentation to the Fire and Aviation Staff of the US Forest Service in June 1991. His insights are equally valid and applicable to lay interactions with fire:

> *I have been asked to give input on wildland firefighter safety. First, let me say I am honored to be able to contribute at this level. The afternoon of June 26, 1990, as I knelt beside a dead Perryville firefighter, I made a promise to the best of my ability to help end the needless fatalities, and alleviate the near misses, by focusing on training and operations pertinent to these goals.*
>
> *Throughout my career I have dealt with wildland fire suppression, as a Hotshot Crew Supervisor, with only minor injuries occurring to those I have directly supervised. This is primarily because of two reasons, luck (which cannot be ignored) and basic lessons which I learned from the exceptional firefighters I have had the opportunity to work with. Many of the really valuable suppression lessons I learned were prior to fire shelter requirements.*
>
> *John Dill, head rock-climbing rescue ranger in Yosemite National Park, recently made an analysis of errors in judgment made preceding an accident. He found three reasons which contribute to accidents; ignorance, casualness and distraction. After thinking about accidents within the firefighter's environment, these same reasons were found to correspond. Allow me to take a moment and help draw the correlations.*

Ignorance: Unfortunately, we have firefighters and fireline supervisors who end up in wildland fire situations that call for skills and knowledge beyond their level of training. I know it is stressed over and over, but the BASICS, basic wildland fire behavior, basic suppression skills, need to be learned and reviewed. Yet many of the entrapments are the result of no lookouts or an insufficient safety zone—a lack of basics.

Casualness: The rock climber standing at the base of a couple thousand-foot granite walls in Yosemite is reassured in his decision to undertake a challenging ascent because of the helicopter that is poised less than a mile from the proposed ascent. We are doing the same. The situation is viewed more casually because we have an option if the tactic fails—our fire shelter.

Another way casualness enters our environment is through the reinforcement of improper tactics since the fire does not "blowup" while we are working the fireline the first few, or several, times. But then we find ourselves entrapped because the familiar situation changes and our reliance on improper tactics just doesn't work this time.

Distraction: Often I have been told that was it not for the on-the-job training that was given by a Division Supervisor, the hazard would not have been noticed and tactics would not have been adjusted. Distraction is a very, very real problem for firefighters. Fatigue and carbon monoxide do not help with the decision-making process, either. Fireline personnel should be continually monitoring each other and remain open to communication and other personnel's evaluation of the situation at hand.

Firefighters still get injured and die in wildfires and structure fires. They still get trapped. Ignorance, casualness and distractions are common threads in all fires that destroy, injure and kill. Rules exist for a reason. Whether they're the 10 & 18, the International Fire Code, instructions from Uncle Bob or advice from the Lnu'k Bear, those rules were borne of ash and blood. We must respect the rules or expect to suffer a similar fate.

7

Sacred Fire

*There the angel of the LORD appeared to him in flames of fire
from within a bush. Moses saw that though the bush was on fire
it did not burn up.*

Exodus 3:2, King James Version of *The Bible*

*As the quintessential immaterial material, that which itself lacks
substance or stable form and yet catalyzes the transformation
of material objects, fire also represents those things that are beyond the material, in the realm of the infinite, including divinity,
the human spirit, and, not least, the challenge to remain dynamic and passionate, to be open to change.*

Rabbi Rob Cabelli

For many children, their first experience with fire comes from candles. We decorate yummy cakes with candles to celebrate another
year of life. We light candles in our homes, even when the electricity is
flowing freely, to add ambiance. Their small flames dance solemnly on
altars and shrines at places of worship.

Unfortunately, representations and use of sacred fire can cause profane blazes in homes, churches and other settings. Most Zen temples
in Japan have burned to the ground at least once, according to author
Colleen Morton Busch, because fire is used for light and cooking within
those ancient wooden structures. She notes that the gate into Eiheiji
monastery is flanked by 600-year old cedars, but the structures within
the gate are less than half that age.

Closer to home, candles are linked to approximately 23,600 residential fires in the United States annually, according to the U.S. Fire Administration's National Fire Data Center. Each year those fires injure 1,525 civilians, kill 165 and cause $390 million in direct property loss. Predictably, candle fires most often occur in December because the candles are placed too close to combustible materials and because they are allowed to burn even after people have left the room. These fires are completely preventable.

In theory, my experience with fire began as a newborn although I doubt I was aware of it at the time. My Mom and I were spending our first Christmas in St. Anthony's Hospital in Denver and "we" wanted a Christmas tree. My Dad went to a local Scandinavian store and purchased a wooden candelabra shaped like a tree that holds nine candles. The Sisters at St. Anthony's gave permission to my parents to burn its candles in lieu of other traditions less suitable for a maternity ward. That candle holder is sacred to me. We continue to burn candles in it on each of my birthdays.

It's not the safest candle holder; its aging, painted wood was bone dry decades ago as it sat on the store shelf. Each year it collects a layer of melted wax. It's also a burn-maker featuring nine points of pain for people who aren't careful. It didn't take me long to realize that lighting the top candle before the two lower tiers was the safest strategy to prevent convective heat burns.

In hindsight, my parents used it to model safe candle behavior. They always placed it on a plate to prevent the wax from ruining the wood table and to create a nonflammable barrier between its candles and tablecloth. They placed it in the middle of the table away from outstretching, curious hands and my sister's (and now my daughters') long locks. They didn't let me light the candles until fourth or fifth grade, and I didn't solo until I was a teen.

That was a much better track record than the interactions I had with candles at our church. The annual candlelight portion of the Christmas Eve service always included hot wax dripping onto my fingers and hand on its way to the carpet. It was always painful at first, then fun in a childish masochistic way and an apparent rite of passage. That more people don't drop their candles onto well-oiled pews and thick carpeting—or sue their pastors for exposures to burns—is noteworthy.

Light of the World

Candles are part of many religious traditions around the world because they symbolically represent both the Divine and divine power. Followers of those religions have integrated "safe" candle fire into their ceremonies—in buildings or shrines in their homes—for centuries. Such sacred fire is rarely threatening, but without understanding the potential energy flickering at the end of the wick, we're setting a stage for failure.

Candles are sacred to Christians because they represent Jesus Christ. As a single candle shines brightly in a darkened sanctuary, Jesus is believed to have brought light and goodness into a dark and oppressive world. "I am the Light of the world," Jesus is recorded as stating in the Book of John, Chapter 8, verse 12, "he who follows Me will not walk in the darkness, but will have the Light of life." But the symbolism runs deeper. When two candles appear on an altar, one on each side, they represent the two natures of Christ: divine and human.

The Christian tradition of Christmas candlelight services, which my family usually attended both in Idaho Springs at Zion Lutheran and at Silver Plume's non-denominational service, recalled two parts of Jesus Christ's legacy. On Christmas Day all those years ago, Jesus was born and brought sacred light to the world. The tradition of passing fire from one candle to the next represents the proselytizing aspect of Christianity: sharing Jesus's light with neighbors and strangers both in the pew beside you and beyond the walls of the church.

The hymn *This Little Light of Mine* elaborates on what that flickering light represents:

> *This little light of mine, I'm gonna let it shine.*
> *This little light of mine,*
> *I'm gonna let it shine, let it shine, let it shine, let it shine.*
> *Won't let Satan blow it out.*
> *I'm gonna let it shine.*

Lutherans use candles, but not as much as their Catholic brethren. For Catholics, a Paschal candle represents Jesus while smaller candles represent individual followers. As long as the candle burns, the disciple is said to be following Christ's example. With violent symbolism, a candle is snuffed out when a parishioner is excommunicated from the church.

The symbolism of candles is amplified in Catholicism. Simply lighting a candle destined to burn at a statue's base isn't simple at all; lighting those candles is a ritual. The faithful Catholic strikes a match and ignites a long stick that passes the flame to the yearning wick of the candle. Votive candles represent sacrifice as the wax vaporizes and burns along the wick. Light from the candle symbolizes going into the light of God and remaining in that condition while returning to a daily routine. Letting the candle consume itself beside a statue symbolizes a Catholic's love of God and her own desire to offer sacrifices—including life itself—for the glory of the Catholic God.

These religious traditions are significant barriers against campaigns for candle safety. Secular candles are relatively easy to abandon in favor of electric, flameless candles. The transition of religious candles from actual fire to simulated fire is worthy of schisms.

For Catholics, the Second Vatican Council in the 1960s allowed for adaptation to a modernizing world. Many Catholics have since embraced modernization for efficiency and practicality, while others abhor it. Does "killing the candle kill the symbolism," as some opponents declare? It may be a philosophical conundrum, but eliminating live fire from buildings does remove one ignition source and reduces the risk of dying in a fire. Ultimately a symbol carries as much symbolism as the beholder grants it.

Candles also are integral to Judaism. Each temple has a light above the Ark in which the congregation stores its Torahs. Originally fueled by oil, that light represents enlightenment, both divine and worldly. Modern Jews also light candles for Shabbat, their holy day of rest between Friday evening and Saturday evening. The most familiar Jewish use of fire is on a menorah. This seven- or nine-branched candelabrum is an ancient symbol for Jews. Seven or nine burning wicks in an otherwise dark tent under a star-filled Sinai sky would have generated considerable illumination: physical and spiritual.

As recorded in Chapter 25 of the Book of Exodus, the Jewish G_d directed Moses to fabricate a seven-branched menorah to represent divine light and heavenly things:

31 And thou shalt make a candlestick of pure gold: of beaten work shall the candlestick be made: his shaft, and his branches, his bowls, his knops, and his flowers, shall be of the same.

32 And six branches shall come out of the sides of it; three branches of the candlestick out of the one side, and three branches of the candlestick out of the other side:

33 Three bowls made like unto almonds, with a knop and a flower in one branch; and three bowls made like almonds in the other branch, with a knop and a flower: so in the six branches that come out of the candlestick.

34 And in the candlesticks shall be four bowls made like unto almonds, with their knops and their flowers.

35 And there shall be a knop under two branches of the same, and a knop under two branches of the same, and a knop under two branches of the same, according to the six branches that proceed out of the candlestick.

36 Their knops and their branches shall be of the same: all it shall be one beaten work of pure gold.

37 And thou shalt make the seven lamps thereof: and they shall light the lamps thereof, that they may give light over against it.

38 And the tongs thereof, and the snuff dishes thereof, shall be of pure gold.

39 Of a talent of pure gold shall he make it, with all these vessels.

40 And look that thou make them after their pattern, which was shewed thee in the mount.

Seven branches reflect the six days of creation and single day of rest. Jews display this menorah throughout the week. It remains a powerful symbol of G_d's presence in the world as noted in the Book of Zechariah: "Not by might, nor by power, but by My spirit."

The nine-branched menorah, called a chanukiyyah, is used during

Chanukah. It commemorates yet another trial in the history of Jews in which they were persecuted by a larger culture. The main temple was destroyed, but the temple's eternal light—a lantern that burned constantly to represent the divine's presence—miraculously burned for eight days with a reservoir of oil that typically lasted only a single day. The ninth branch of this chanukiyyah, which is usually taller, shorter or otherwise different from the others, represents the shamash. He is responsible for lighting and tending to the menorah.

Although fire is considered sacred, representing both the divine and human spirits, building a fire is considered work. Therefore Jews are unable to build fires during their Sabbath because work is forbidden.

A Spiritual Sanitizer

For many versions of Christianity and Judaism, fire also serves as a spiritual sanitizer, violently revitalizing physical and spiritual landscapes: "A fire devoureth before them; and behind them a flame burneth: the land is as the garden of Eden before them, and behind them a desolate wilderness; yea, and nothing shall escape them" and later, bringing the Lord's wrath against Israel for ignoring His statutes: "Then take of them again, and cast them into the midst of the fire, and burn them in the fire; for thereof shall a fire come forth into all the house of Israel."

According to the Book of Mormon, the High Priest Alma also referred to fire as a sanitizer, destroying deviants in society:

> And again I say unto you, the Spirit saith: Behold, the ax is laid at the root of the tree; therefore every tree that bringeth not forth good fruit shall be hewn down and cast into the fire, yea, a fire which cannot be consumed, even an unquenchable fire. Behold, and remember, the Holy One hath spoken it.

Jews and Christians initially learned to worship fire in the Old Testament Book of Exodus when an Angel of the Lord speaks to Moses from a burning acacia bush.

> *1 Now Moses was tending the flock of Jethro his father-in-law, the priest of Midian, and he led the flock to the far side of the wilderness and came to Horeb, the mountain of God.*
> *2 There the angel of the LORD appeared to him in flames of*

fire from within a bush. Moses saw that though the bush was on fire it did not burn up.

3 So Moses thought, "I will go over and see this strange sight—why the bush does not burn up."

4 When the LORD saw that he had gone over to look, God called to him from within the bush, "Moses! Moses!" And Moses said, "Here I am."

5 "Do not come any closer," God said. "Take off your sandals, for the place where you are standing is holy ground."

6 Then he said, "I am the God of your father, [a] the God of Abraham, the God of Isaac and the God of Jacob." At this, Moses hid his face, because he was afraid to look at God.

It's interesting that the Old Testament God chose to appear as fire, a potentially destructive phenomenon, yet this fire didn't even singe the bush.

The story is told in the Quran (the Book of TaHa) as well:

9 And has there come to you the story of Musa [Moses]?

10 When he saw a fire, he said to his family: "Wait! Verily, I have seen a fire, perhaps I can bring you some burning brand therefrom, or find some guidance at the fire."

11 And when he came to it [the fire], he was called by name: "O Musa!

12 Verily! I am your Lord! So take off your shoes, you are in the sacred valley, Tuwa.

13 And I have chosen you. So listen to that which is inspired to you.

It's not surprising that Moses was mystified why the bush wasn't burning because the thorny wild acacia is dry and brittle. Muslims would suggest the bush didn't burn because Allah ordered the fire to protect that bush. Several modern branches of Christianity refer to the unconsumed burning bush in their logos and mottos as representing resilience in the face of struggle.

For the five member cultures of the Iroquois Nation, fire symbolizes another sacred entity: government. According to Gayanashagowa, the

Great Binding Law of the Iroquois Nation, a burning fire represented resilience, justice and life:

Article 3. To you Adodarhoh [the creator], the Onondaga cousin Lords, I and the other Confederate Lords have entrusted the caretaking and the watching of the Five Nations Council Fire.

When there is any business to be transacted and the Confederate Council is not in session, a messenger shall be dispatched either to Adodarhoh, Hononwirehtonh or Skanawatih, Fire Keepers, or to their War Chiefs with a full statement of the case desired to be considered. Then shall Adodarhoh call his cousin [associate] Lords together and consider whether or not the case is of sufficient importance to demand the attention of the Confederate Council. If so, Adodarhoh shall dispatch messengers to summon all the Confederate Lords to assemble beneath the Tree of the Long Leaves.

When the Lords are assembled the Council Fire shall be kindled, but not with chestnut wood, and Adodarhoh shall formally open the Council.

Then shall Adodarhoh and his cousin Lords, the Fire Keepers, announce the subject for discussion.

The Smoke of the Confederate Council Fire shall ever ascend and pierce the sky so that other nations who may be allies may see the Council Fire of the Great Peace.

Adodarhoh and his cousin Lords are entrusted with the Keeping of the Council Fire.

Article 4. You, Adodarhoh, and your thirteen cousin Lords, shall faithfully keep the space about the Council Fire clean and you shall allow neither dust nor dirt to accumulate. I lay a Long Wing before you as a broom. As a weapon against a crawling creature I lay a staff with you so that you may thrust it away from the Council Fire. If you fail to cast it out then call the rest of the United Lords to your aid.

Bridging Sacred and Secular

Many other cultures utilize fire to connect the worldly with the other-worldly. Fire is a complex topic in the Pre-Islamic Iranian tradition. It is both an earthly representative of the divine as well as a deity itself. While individual families maintain a hearth, they also venerate the Bahram fire that burns in a special fire temple. Creating this sacred fire lasts an entire year and then a priest called a mobed tends to it five times a day.

> *The lower part of his face is covered with a veil [Avesta pait-idana], preventing his breath from polluting the sacred fire, and his hands are gloved. He lays down a log of sandalwood and recites three times the words* dushmata, duzhukhta, duzh-varshta *to repel "evil thoughts, evil words, evil deeds."*

This fire also is known as Berezisavanh. It appears before Ahura Mazda, the Supreme Creator and Wise Lord. In this tradition documented in the *Avesta Yasna*, Iranians identify four other types of fire: Vohu Fryana, which burns in the bodies of men and animals, keeping them warm; Urvazishta, which burns in the plants and can produce flames by friction; Vazishta or lightning that purifies the sky and slays the demon Spenjaghrya; and Spenishta, the most holy type that burns in paradise in the presence of Ahura Mazda.

While many cultures don't address how such a sacred creation could harm humans, Iranians blamed Angra Mainyu, the destroyer. This deity already had covered useful plants with bark and thorns or added poison to their sap to threaten humans. In the case of fire, Angra Mainyu gave it smoke and darkness to hide its potential for creation and illumination.

Zarathushtra, a prophet in ancient Iran, viewed fire—called Athra—as a symbol of Ahura Mazda. Athra guides and protects the faithful along the path of Truth (Asha):

> *Thus, O Ahura we yearn through Asha for thy Fire, mighty, most-enduring and courageous, giving clear guidance in life to the earnest believers; but O Mazda to those with destructive tendencies, it overcomes their violence by the power of its flames.*

Zarathushtra and his followers, the Zoroastrians, don't worship fire by itself; they venerate it as a representation of Ahura Mazda and use it to experience the light and warmth of Ahura Mazda's presence.

Within the Islamic culture, fire is a creation of Allah for which Muslims are grateful. Like other creations, it harms or benefits people only on Allah's orders. Prophet Ibrahim (Abraham) was thrown into a fire after questioning the worship of other deities and then destroying representations of them, as described in the following excerpt from the Quran, and he escaped unharmed because Allah ordered that fire to cool.

> 67 *He said, 'Do you then worship instead of Allah that which cannot profit you at all, nor harm you?*
>
> 68 *'Fie on you and on that which you worship instead of Allah! Will you not then understand?'*
>
> 69 *They said, 'Burn him and help your gods, if at all you mean to do anything.'*
>
> 70 *We said, 'O fire, be thou cold and a means of safety for Abraham!'*

Assistant Imam Fudail Hassan, whom I interviewed at a local mosque, explains that fire helps many aspects of life for Muslims, but it's a "severe enemy" if not used with care. The Prophet Mohammed instructed Muslims to guard and protect themselves "by treating fire well." Muslims are ordered to extinguish fires and turn off heat sources before sleeping.

Fire also is an essential part of the Islamic Hell, as described in the 24th verse from the Book of Al-Baqarah: "Fire is part of Hell that burns men and stones, and is home to people who do not believe in the ways of Islam." Hellfire serves a pair of purposes. These sacred fires punish people who have received but deny the teachings of Allah, and purify believers who have sinned but have not yet repented.

> 7 *Thus has the Word of your Lord been justified against those who disbelieved, that they will be the dwellers of the Fire.*
>
> 8 *Those [angels] who bear the Throne [of Allah] and those around it glorify the praises of their Lord, and believe in Him, and ask forgiveness for those who believe [in the Oneness of Al-lah] [saying]: "Our Lord! You comprehend all things in mercy*

*and knowledge, so forgive those who repent and follow Your
Way, and save them from the torment of the blazing Fire!*

Since Muslims are taught patience, gratefully enduring this second
opportunity for paradise seems a reasonable test.

Latter Day Saints interpret the destination of those who reject the
words of the prophets, Christ, Holy Ghost, Holy Spirit and God similarly. Within the sixth chapter of the Book of Jacob we find the warning:

*And according to the power of justice, for justice cannot be
denied, ye must go away into that lake of fire and brimstone,
whose flames are unquenchable, and whose smoke ascendeth
up forever and ever, which lake of fire and brimstone is endless
torment.*

Joseph Smith received and shared similar revelations in the 1830s
that are now found in the Doctrine and Covenants of the LDS:

*And the righteous shall be gathered on my right hand unto
eternal life; and the wicked on my left hand will I be ashamed
to own before the Father; Wherefore I will say unto them—De-
part from me, ye cursed, into everlasting fire, prepared for the
devil and his angels.*

Not all Christians share that interpretation of Hell. Jehovah's
Witnesses don't believe Hell is hot or the site of permanent torment.
While all-consuming fire represents Jehovah God and His ability to
accomplish tasks, Jehovah God is love and would not use fire to torment followers for disbelief. Instead, Hell is devoid of life because it is
synonymous with the grave or death, based on its etymological roots.
Although supported by verses in *The Bible*, their interpretation is not
held widely.

The worship of fire was so deeply rooted in Armenian culture
that Christian authors somewhat hypocritically referred to heathen
Armenians as ash-worshippers. Fire was an earthly manifestation of
the sun and lightning. Fire gave heat and light. Even today to extinguish
a candle or a fire is not a simple matter among traditional Armenians.
It requires care and respect, according to author Mardiros Ananikian.

Fire must not be desecrated by the presence of a dead body, by human breath, by spitting into it, or burning unclean things such as hair. An impure fire must be rejected and a purer one kindled in its place, usually from a flint.

The following Alabama-Coushatta story reveals a similar reverence regarding the creation of fires.

> *...The Fire nearly went out and called aloud. It was almost extinguished. "Feed me," it said. Then some human beings saw it. They got a stick toward the north and laid it down upon it. They got another stick toward the west and laid it down upon it. They got a stick at the south and laid it down there. They got another at the east and laid it down and the Fire blazed up.*

The feeding of wood into a fire is a sacred process for both the Alabama and Coushatta people, who were from what is now the southeastern United States. Although these tribes were separate, they shared common religious beliefs. According to researcher R. E. Moore, fire was an essential part of their spirituality. Each home, and the village as a whole, had fires burning as much as possible. To build the fire, they would:

> *Place four logs in the shape of a cross around the central fire. One log would point north, one east, one south and one west. As the fire burned the ends of these logs the people would push them in to the center. A home fire would have small logs and a dance ground would have big logs to last longer. Fire was believed to be a part of the sun and the sun represented the highest God. The Southeastern Indian symbol for fire and the sun looks like a cross with four arms, each representing the sacred logs, radiating from a small circle in the middle, representing sacred fire.*

Fire was sacred to other inhabitants of the western hemisphere as well. There were rules to follow and consequences for violating them, as Rabbit discovered in the following Creek story:

> *All the people came together and said: "How shall we obtain fire?" It was agreed that Rabbit should obtain fire for the people.*

He went across the great water to the east. He was received gladly, and a great dance was arranged. Then Rabbit entered the dancing circle, gaily dressed, and wearing a peculiar cap on his head into which he had stuck four sticks of rosin.

As the people danced they approached nearer and nearer the sacred fire in the center of the circle. The Rabbit also danced nearer and nearer the fire. The dancers began to bow to the sacred fire, lower and lower. Rabbit also bowed to the fire, lower and lower. Suddenly, as he bowed very low, the sticks of rosin caught fire and his head was a blaze of flame.

The people were amazed at the impious stranger who had dared to touch the sacred fire. They ran at him in anger, and away ran Rabbit, the people pursuing him. He ran to the great water and plunged in, while the people stopped on the shore.

Rabbit swam across the great water, with the flames blazing from his cap. He returned to his people, who thus obtained fire from the east.

In the following Zia story, it's no accident that Coyote has to pass four guards at four doors to reach the sacred fire. Four is a number sacred to the Zia people of New Mexico. Their image of the sun, which resides on the state flag of New Mexico, consists of a red circle with rays pointing in four directions. Its power is revealed in the four seasons and four sacred obligations that each Zia must develop: strong body, clear mind, pure spirit and devotion to the welfare of others.

A long, long time ago, the people became tired of feeding on grass, like deer and wild animals, and they talked together how fire might be found. The Ti-amoni said, "Coyote is the best man to steal fire from the world below," so he sent for Coyote.

When Coyote came, the Ti-amoni said, "The people wish for fire. We are tired of feeding on grass. You must go to the world below and bring the fire."

Coyote said, "It is well, father. I will go."

So Coyote slipped stealthily to the house of Sussistinnako.

It was the middle of the night. Snake, who guarded the first door, was asleep, and he slipped quickly and quietly by.

Cougar, who guarded the second door, was asleep, and Coyote slipped by. Bear, who guarded the third door, was also sleeping. At the fourth door, Coyote found the guardian of the fire asleep. Slipping through into the room of Sussistinnako he found him also sleeping.

Coyote quickly lighted the cedar brand which was attached to his tail and hurried out. Spider awoke, just enough to know someone was leaving the room.

"Who is there?" he cried. Then he called, "Someone has been here." But before he could awaken the sleeping Bear and Cougar and Snake, Coyote had almost reached the upper world.

In accepting his task to steal fire, Coyote had devoted himself to the welfare of the Zia even while using his trickster spirit to do so through theft rather than a less treacherous tactic.

Sharing Fire

Because fire was and remains essential for survival, it has immense value. That value can make one person wealthy, but sharing fire with others expands its sacred and secular value exponentially. It's in the sharing of fire that it becomes and remains sacred. In my experience, sitting next to a campfire or the fireplace alone is fine, but sharing the experience with friends and loved ones is preferred.

I had hoped our first family campfire would have been more intimate or at least romantic, but a nearing cold front put the kibosh on that plan. Instead, our first campfire along a plateau above the northern shore of Pueblo Reservoir was a chilly, short-lived moment. We built the fire using newspaper, kindling, and firewood from home. All four of us sat around the windward side of the fire in our appropriately-sized camp chairs to prevent smoke and embers from showering us. My daughters Kaliska and Tabor seemed to enjoy the idea, but were chilly enough in their pajamas to prefer the comfort of the pop-up camper. The evening reminded me of ill-fated campfires on the windswept red sandstone in Moonshine Wash and the gusty east shore of Carter Lake. The four of us enjoyed other campfires during the summer of 2012, but as the first one, I hope to keep it in my memory. ✎

8

Risk Perception and Fire

When we know something and rest in that knowing, we limit our vision. We will only see what our knowing will allow us to see. In this way experience can be our enemy.

Zoketsu Norman Fisher, founder and teacher of the Everyday Zen Foundation

Fire is simply fire. It has no sense of morality, no persona, does not wish to do good or bad, is neither deliberately enemy nor friend.

Douglas Gantenbein

*T*he Retreat is a small subdivision within the City of Castle Pines in Douglas County, Colorado. Castle Pines is named for the towering ponderosa pines that grow in the area, which is in the northern part of the Black Forest ecosystem. Those pines grew tall and old because wildfires periodically burned there, removing competition for scarce resources and recycling nutrients into the soil. Periodic fires also created conditions ripe for Gambel oak. The wood-frame homes in The Retreat surround four cul-de-sacs and they, in turn, are adjacent to open space managed by the homeowners association. The oak and grasses have grown largely unchecked since the subdivision was developed in the 1980s. Simultaneously, landscapers and homeowners planted junipers, Austrian pines, fitzers and other flammable vegetation around their homes.

Because wildfire may return and threaten both the Retreat and the rest of Castle Pines, its open space was noted as one of four priority wildfire mitigation projects in the city's community wildfire protection plan. As far back as 2006, South Metro Fire Rescue educated homeowners about the risks wildfire posed to their homes. We increased our efforts in 2012 applying for a grant on behalf of the homeowners association to mitigate oak on ten acres of its open space. Not surprisingly, the news that we had secured funds to perform the mitigation was not met with as many open arms as we had hoped. Instead of open arms and jubilation we encountered crossed arms and cross words from homeowners scared that we planned to clear-cut all their oak—including oak growing on private property—and make their neighborhood into an ugly wasteland. When asked what changes they wanted from mitigation efforts, their answers resembled these two comments:

- *Any possible changes or alterations to the open space should be minimal if done at all. My preference is to leave the open space in its absolute NATURAL state. That's why we chose to live here.*

- *Nothing!!! People move here because of the natural landscape, then want to chop into it to supposedly correct it.*

Even after a two-hour community meeting, many residents held onto their opinions that mitigation was wrong, misplaced and unnecessary because "there's never been a fire in the Retreat," as one resident emphatically told me.

Enter the black swan. Back in the 17th century, when explorers were cruising the seven seas encountering unfamiliar lands, animals, and people while searching for resources, one group found a bird in Australia that looked like a swan in all ways except that it was black. As good scientists, they killed it and brought it back to the Royal Academy where the brightest minds agreed the bird couldn't be a swan because all swans were white.

They were wrong, but this incident created a useful concept. As humans, most of our knowledge is based on observation and experience. If we haven't observed it or experienced it, we tend to doubt that it—whatever it may represent—could exist or occur, as author Nicholas Taleb explains in his book, *The Black Swan*:

> *Consider a turkey that is fed every day. Every single feeding will firm up the bird's belief that it is a general rule of life to be*

fed every day by friendly members of the human race. On the afternoon of the Wednesday before Thanksgiving, something unexpected will happen to the turkey.

Projecting the future from the past, he argues effectively, is naïve and dangerous. Modern black swans are events that occur without recorded precedent or in frequency or severity that violates experience. Clearly, point of view matters. A wildfire burning through the Retreat would be a black swan to many homeowners, but not to me, our firefighters or the foresters who have contributed to the discussions there.

Ultimately, we in the mitigation and prevention worlds confront such reluctance to prepare for black swans on a daily basis. We frame the argument around "when"; our audiences often prefer "if." The only proof of our success is when a swan lands, as it did in the town of Alpine, Arizona, in 2011.

On the third day of June in 2011, the Wallow Fire—a 40,000-acre wildfire—sprinted toward the town of Alpine through tree canopies in a crown fire. Embers landed a mile away igniting spot fires across the valley. Driven by high winds and enhanced by low humidity, the crown fire expanded unchecked through the forest, crested a ridge and burned downhill toward town. As it approached town, it encountered the White Mountain Stewardship Fuel Treatment project, which radically changed the fire's behavior.

That fuel treatment project, started in 2004, had thinned trees to create space between remaining tree canopies, cut low-lying branches that are called ladder fuels because they let fire climb into canopies, and removed brush to eliminate another rung in the vertical fuel ladder. The effort is summarized in the "How Fuel Treatments Saved Homes in the 2011 Wallow Fire" report produced by the Wildland Fire Lessons Learned Center. Homeowners also created defensible space around their homes and hardened structures to resist ignition.

As the raging crown fire reached the treated land, it dropped from the tree crowns onto the surface fuels and the rate of spread rapidly decelerated. Flame lengths shortened enough that firefighters could engage flames directly. Firefighters extinguished spot fires and conducted burnout operations to eliminate unburned fuels between the main fire and homes still at risk. Although a wildfire had not burned through Alpine in recorded memory, residents and land managers cooperated to reduce their risk and prevented all but one structure from burning. Had residents not mitigated around and within their town, "many of the houses would have caught fire and burned to the ground," according to Fire Management Officer Jim Aylor. Had they not mitigated, Alpine would resemble countless other burned-out communities.

Alpine is yet another success story, but such success stories don't sway the naysayers as easily as we would think. Nor does research into the cost of wildfire. The 2013 Rim Fire in California caused at least $1.8 billion in damage, which amounts to approximately $7,800 per acre. Spending even half that amount per acre on mitigation could have created a vastly different outcome.

It's Not Ignorance

A team of U.S. Forest Service sociologists conducted focus groups in five urban interface communities in 2004 to determine whether residents understood their risk for wildfire. They met with residents in Boulder, Colorado; Flagstaff, Arizona; Hamilton, Montana; Reno, Nevada; and San Bernardino, California. Each of the cities was familiar with wildfires and part, if not all, of each jurisdiction could be considered wildland urban interface (WUI).

The researchers found that residents were aware of their wildfire risk generally, but tolerated that risk better than people who didn't live there. Residents also placed more value on recreation, privacy, aesthetics and other qualities than on mitigation. Their scales tipped away from wildfire mitigation, which created a significant barrier between them and the local subject-matter experts who weighted mitigation more heavily than the other attributes.

Those are important discoveries. Ignorance of fire's risk is present in our communities, but it isn't prevalent.

Research into natural hazards has shown that people act to reduce risk more often for hazards that are more pervasive in their effect: long duration, large area affected, and/or more frequent. They are less likely to act for hazards with a quick onset, short duration and/or small area. The latter hazards are forgotten quickly—usually within three to six months. Wildfires clearly fit in the second category.

Risk tolerance is important. According to the USFS focus groups, people who lived near a wildfire-prone ecosystem consistently rated the wildfire risk higher than those who lived within a wildfire prone ecosystem or interface. WUI dwellers tolerate more wildfire risk than folk who live outside the vicinity. Not surprisingly, researchers linked lower risk perception to higher benefit perception. When those residents chose to live in a WUI setting, they also accepted the associated costs, including wildfire risk.

Comments from residents of The Retreat reflect the researchers' revelations:

- *Minimal fire mitigation makes sense. To change the appearance of the open space on the chance that we could have a fire doesn't make sense to me.*

- *South Metro has a fire station one minute from the Retreat. We are not a mountainous community, or placed in the middle of a pine forest. In my opinion the funds FEMA has offered should be used in a more meaningful way.*

- *Given that I think the wildfire risk is low, I would rather not change anything from the way it is today. We are not dense like the Black Forest, or the Waldo Canyon area that burned within the past 3 years.*

- *Fires cause major damage where firefighting has limited access. The Retreat and Castle Pines has easy access to fighting fires, I do not perceive any danger here.*

While each of those opinions differs greatly from ours as wildland fire mitigation specialists, each of them also is valid. They are entitled to balance costs and benefits based on their own values.

My family did. We lived in Alice, an unincorporated village at 10,000 feet above sea level in northern Clear Creek County, Colorado, for eight years. We accepted the harsh winter days, long commutes, delayed ambulance service, power outages, and wildfire risk because they were overshadowed by sun-drenched afternoons—even in the middle of January, privacy, belonging to a small rural fire department, amazing wildlife, and close access to backcountry recreation. We created the best defensible space in the neighborhood knowing we, as firefighters, should model the right behavior and would probably be too busy helping neighbors evacuate to the last-chance meadow rather than engaging the fire when the big one ignited in our lodgepole- and Douglas fir-filled subalpine forest.

Sarah McCaffrey, one of the sociologists, suggested a new strategy for addressing wildfire risk, a suggestion that is applicable to other types of fires as well: "Rather than focusing on raising risk perceptions... focus on changing the perceived balance of risk and benefits." In other words, decreasing risk could add to the benefits of living in a fire-prone environment. For example, reducing the volume of scrub oak can create new trails for recreation and diversify habitat for critters. Similarly, adding residential sprinklers provides peace of mind while living miles away from the nearest fire department.

When I surveyed residents of The Retreat for what they liked about their neighborhood, they answered with the same replies that McCaffrey received:

- *I love that we back up to open space and have privacy, due to the trees.*

- *Unfenced and easy access to the open space. It feels like having some semblance of the wilderness right in our backyard. The natural surroundings are a very pleasant barrier to traffic noise and other outside disturbances. My backyard is a sanctuary. I love seeing the wildlife right outside my backdoor.*

- *I like that it backs up to the open space that has deer, rabbits and bear.*

- *I like the privacy that the scrub oak provides.*

If we would have emphasized their values and made wildfire mitigation secondary, we probably would have found greater success. It's an economic argument. If perceived costs outweigh perceived benefits of preparing for an unlikely event, it's a rare person who will act. It's important to remember that costs are more than fiscal. Regarding The Retreat project, homeowners would have been responsible for only $570 collectively but their costs included perceived losses in aesthetics, privacy and sound barriers. They saw an expensive project where I saw a thrifty one. Because their values counted at least as much as ours, we respected a community vote and returned the $9,000 grant to FEMA.

Waiting for the Cavalry

In 2008 the Basin Complex Fire burned through the Los Padres National Forest in central California. It famously threatened and burned through the Tassajara Zen Mountain Center in July in which five monks had ignored official orders to evacuate and successfully defended its structures. Colleen Morton Busch shared the monks' experience in her book *Fire Monks*. Risk perception is a theme that runs through the book. As the wildfire neared the monastery, pushed by extremely hot, dry and windy conditions, several firefighters advised and later instructed everyone to evacuate. When the five ignored the evacuation order, they did so partially because they underestimated their risk and overestimated the willingness of firefighters to respond. They expected firefighters, who perceived the wildfire from a different set of values and were governed by a different set of rules, to push aside those values and rules to sacrifice their own safety to save five adults who repeatedly ignored evacuation orders.

The monks weren't alone in that assumption. Busch included several comments from former Tassajara residents and current allies—some of whom were firefighters—complaining that the "cavalry" didn't ride into the valley to save those monks and property. One former resident who had survived the 1977 Marble Cone Fire's run through Tassajara blogged for supporters to call their elected officials and make fire departments send resources into the valley, adding "their presence makes it much more likely that further assistance will come from CAL FIRE." This one-sided discussion, captured in the book, makes me wonder why the cavalry should be expected to rescue adults who stayed in a vulnerable situation against the best advice from subject-matter experts.

Zen theory is accurate: every act happens only once. No two fires burn exactly the same. While one wildfire may have burned through a heavily vegetated, steep-sloped, wildfire-prone valley and caused certain conditions in 1977, it's delusional to believe a different wildfire would behave exactly the same. Firefighters also have new rules of engagement so that more of them can go home safely at the end of their shifts. The cavalry did arrive in 1977, but that act was over.

The American fire service does bear some responsibility for this attitude. We portray ourselves as the cavalry, as heroes who go in when everybody else goes out, to slay dragons and fight what others fear. In the process, we create expectations that firefighters will always save the day and no one else has any responsibility to protect themselves aside from buying an insurance policy.

As the last decades of fatal fires have shown, that's flawed thinking. While that expectation is fair during normal operations, during large-scale emergencies, citizens may be disappointed. There aren't enough firefighters to go around. There aren't enough crews to put a fire engine in each driveway. Firefighter safety is paramount and more important than saving homes.

Residents question and fight against mitigation projects. Developers question and fight against sprinkler systems. Adults question and fight against evacuation drills.

But it's time to ante up. ✎

9

Harmony with Fire

All peoples live in a physical world which is not only natural but also historical—a creation of their ancestors and themselves.

Richard White, *Land Use, Environment, and Social Change: The Shaping of Island County, Washington*

A mere unquestioned acceptance of an unproven assumption does not constitute proof.

James C. Malin, *History and Ecology: Studies of the Grassland*

*B*ecause fire is sacred to many people, trying to ignore it, fight with it, or manipulate it won't work. Seeking harmony with it, respecting it and renewing a healthy relationship are essential for our safety and well-being. Wildfire is an ideal place to begin.

Since I joined the fire service in 1998, I've become familiar with the annual predictions of wildfire seasons. Colorado's traditional wildfire season was summer. Each time a warm, dry spring encountered summer we in the fire service would postulate about the severity of the year's probable wildfires. Other firefighters no doubt undertake similar discussions, blending experience with predictions of weather and fuel conditions, regardless of their burning seasons.

Yet wildfire season is a misnomer; ultimately, Colorado's wildfire season stretches from January 1 to December 31 whenever there is no snow on the ground, a source of heat, and too often a moment of naïve carelessness.

The year 2012 held a special place in the memories of Coloradans; it was the 10-year anniversary of the summer that "all of Colorado was on fire." Then Governor Bill Owens made that exaggerated comment as dozens of smoke plumes clouded the state's usually crisp and clean air and painted its sunsets an eerie orange starting in the spring with Schoonover, Snaking and Black Mountain, and extending through the summer with Big Elk, Missionary Ridge, Coal Seam, Hayman and hundreds of others. Yet the area burned was only 382 square miles, which is less than one percent of the state's total land area.

While the fires of 2002 are a distant memory for many Coloradans, and missing entirely from the minds of thousands of residents who were born or moved to the state in the intervening decade, the fires of the 2010s remain viable and visible memories because they burned homes, killed civilians, scorched infrastructure and killed firefighters on the evening news and social media. Each fire that burns into an ecosystem with human homes, whether it's the Bastrop County Complex (Texas 2011), Waldo Canyon (Colorado 2012), Yarnell Hill (Arizona 2013), Carlton Complex (Washington 2014), or thousands of others, briefly reminds residents of wildland urban interface (WUI) neighborhoods that they live in a wildfire-prone ecosystem.

The WUI isn't new. North America has always been an interface of wildland and urban influences as Native American, Norse, Spanish, Dutch, French, Asian, English, African, and Latin American colonists immigrated and settled here. As cultures experienced the ecosystems and encountered one another, they learned new rules about the role of fire in their area. They learned how to use fire, how to avoid misusing it, and how to respond to it. They adapted to the presence of fire with education, engineering, enforcement and economic strategies, much like we do today.

"The Savages are accustomed to set fire of the Country in all places where they come, and to burne it twize a yeare, at the Spring, and the fall of the leafe," Thomas Morton wrote in his 1637 book *New English Canaan*:

> *And least their firing of the Country in this manner should be an occasion of damnifying us, and indaingering our habitations, wee our selves have used carefully about the same times to observe the winds, and fire the grounds about our owne habitations; to prevent the Dammage that might happen by any*

*neglect thereof, if the fire should come neere those howses in
our absence.*

While he only recognized these burning operations as a nuisance
or threat, other colonists in what is now upstate New York appreciated
how native peoples used fire as a tool:

> *Their method of clearing their lands without the use of an
> axe, was less difficult, than we should imagine. When their old
> cultivated fields were worn out, and new ones wanted, they
> selected the spot to be cleared, and kindled fires around all the
> trees which they wished to destroy, and kept them burning till
> the bark was consumed and the trees killed while standing.
> This done, the ground was planted, and the dead trees left to
> perish and fall, without further labour.*
>
> *When the trees were thus fallen, they were burned in pieces
> of such length as that they could be rolled together into piles
> and burnt. In this way the field was in a few years entirely
> cleared. A smart woman among a large number of these fallen
> dry trees, would burn in pieces, as many as an expert axe-man
> could cut in two or three days.*

Clearly, fire use was an important tactic that native peoples employed
to influence their environment purposefully and profoundly.

Native North Americans from coast to coast utilized fire to manipulate their ecosystems. Native peoples in the Sierra Nevada, such as
the Maidu, Miwok, Washoe, and Tubatulabal cultures, burned parts of
their ecosystems to enhance plant production. More oak meant more
acorns. Wildfires also maintained grass and shrub species needed for
basket-making resources. An abundance of bracken and camas on
Whidbey Island and other prairies in the Pacific Northwest suggest
that native peoples such as the Salish and Skagits burned their environments to promote growth of those edible plants. In fact indigenous
communities across the continent also used fire to drive game, from
rabbits to larger ungulates, and to roast insects.

James Cooper recorded the botany of the Pacific Northwest as part
of the railroad survey of the Stevens Railroad Survey between 1853 and
1855. In southwestern Washington Territory, he found young spruce

trees growing rapidly where "this burning had been prevented for twenty years past... and Indians have told me that they can remember when some other prairies were much larger than at present."

Prairie Fires Were Common

American pioneers had to learn old lessons about fire as they ventured beyond the perimeter of established communities into unfamiliar landscapes. Pioneer Lizzie Dopps recalled her experiences with wildfires in early 20th-century rural Kansas. Although embellished with dramatic flair, her recollections capture the awe, fear and acceptance of fires that should sound familiar to 21st-century wildland urban interface residents:

> *Prairie fires were not uncommon. In fact, we lived in constant fear of them during the hot dry season—these monstrous infernos raging and leaping towards us like a thing imbued with life, a demon gone into a wild frenzy.*
>
> *To avoid a prairie fire, we plowed many furrows around our fields and homes, leaving them bare, so that no dried vegetation would feed these furies, but sometimes this did not stop them and we had to fight these demons.*
>
> *One evening after darkness fell there was a rosy glow in the sky. We felt sure it was a prairie fire in the distance.*
>
> *After anxiously watching and waiting, there it came in all its fierce, red fury, making a direct line for our home.*
>
> *[Eli] arose immediately and neighbors came to help. On and on it came like a wild hungry animal let loose, greedy little red tongues darting out, licking up and devouring everything in its way.*
>
> *The men tore up blankets and wet them in pails and tubs of water and went at it with a vengeance. On and on it came, over the fields of stubble. Now it was licking up the loose straw around the straw stack!*
>
> *In the house it was light as day, a red glare coming through the window, although it was in the dead of night.*

Clearly beyond their element, some settlers turned to Native

Americans for lessons about reacting to fires. According to one Texas pioneer, animal guts could extinguish a low-intensity wildfire:

> *Workers killed a cow, slit the carcass up the middle and spread the body on the ground (slimy side down). Next they tied a rope to the front legs and another to the back. Two riders attached their ropes to their saddle horns, and, on horseback, they dragged the heavy dead cow, moist innards and all, along the fire line, smothering the flames as they rode... Other men followed the drag, moving along the fire line on foot with wet blankets or sacks to beat out any leftover hot spots.*

Those cowboys learned from their predecessors how to break the heat leg of the fire triangle. Wet fuel is more fire-resistant than dry fuel.

Defining Where Wildfires Burn

Fire and the wildland urban interface zones in which it burns garner new attention from our modern media and residents each time a wildfire dances through a community and burns homes, scares livestock, and inconveniences people. Sometimes it injures and kills people, too.

A common reaction during and after wildland urban interface fires in rural communities is the non-participant's question, "Why do those people live there?" or the complaint, "Why are those people allowed to live there?" Those questions flooded Colorado media during and after the June 2013 Black Forest Fire. Answers are complex.

In the 1990s, many subject matter experts officially defined the wildland urban interface as residential areas within a quarter mile of the untouched woods. While I appreciate the desire to define a hazard, which makes it less scary and chaotic for our western cultures, this definition is flawed. WUI fires aren't restricted to forested areas and risk isn't confined to the quarter-mile distance. Wildfires that threaten or burn a mix of vegetation and human structures also occur in shrubland and grassland ecosystems. Over 25,000 wildfires charred thousands of acres, incinerated hundreds of homes and other structures, and killed livestock in Texas in 2011. Those fires also killed civilians and firefighters.

Winds can drive embers far beyond a quarter mile in advance of a wildfire. A spot fire on an unrated wood shake-shingle roof, in a desiccated leaf-filled gutter, or a mature "little green gas can" is still a WUI fire.

I once sparred with a fellow *High Country News* reader on the issue of WUI risk reduction. He, a resident of Bozeman, Montana, complained at the lack of laws prohibiting people from living in the wildfire prone ecosystems of the West. I argued that his proposed laws would eliminate Bozeman and other western towns from the map. He replied that most of Bozeman is more than a quarter mile away from the surrounding forest. I let it go, initially, and returned to work but I was seething inside. Embers don't extinguish magically at a quarter mile. Remove the first quarter mile of homes surrounding Bozeman, and another ring of homes would be in the quarter-mile zone. Finally, I considered, at some point all American communities were in the wild-land urban interface.

Indirect Ignitions

Jack Cohen's research has proven scientifically beyond a doubt what firefighters have observed and known intuitively for decades. A wall of flame isn't responsible for igniting structures; the culprits are embers, radiant heat, and low-intensity creeping fires.

Embers—airborne chunks of burning materials also known as fire-brands—sail with the wind and penetrate vents to land on insulation, or on cushions of deck furniture, stacks of firewood, leaf- and needle-filled gutters, debris under decks, dry wood shingles, flammable vegetation such as junipers and piles of dead leaves, and needles along walls. Heat radiates from nearby vegetation—especially junipers, fitzers, mugo pine, piñon pine, cedar and arborvitae—or from other burning structures against wood siding and decking to ignite homes. Poorly maintained lawns, aging cedar fencing, and carpets of desiccated pines needles lead creeping flames directly to homes. Structure-to-structure conflagrations, ignited after embers and creeping fires ignite one home on a block, also count as WUI fires that sprint beyond the quarter-mile boundary theory. The proof for Cohen's argument is prolific.

The old quarter-mile definition also ignored potential exposures surrounding greenbelts and open space within suburban and urban areas where home densities result in significant damage from relatively small wildfires. In August 2010 firefighters and residents of Ashland, Oregon, watched embers from a grass fire on one side of Interstate 5 jump the highway onto the beautiful yet extremely flammable vegetation planted beside homes on the other side. As junipers and arborvitae burned, they radiated immense heat and spewed embers that ignited homes. Structure-to-structure ignitions eventually destroyed 11 homes in under an hour. That fire is but one example of many.

Another unnecessary exercise is distinguishing between interface and intermix communities. State and federal agencies use these categories to discern differences in housing density and vegetation. Interface is land with higher housing density adjacent to natural areas; intermix occurs where homes "mingle" with natural areas, according to the US Forest Service. Those definitions may be useful for grantors, but they obscure the reality of wildfire in a haze of terminology.

The NFPA updated its definition of the WUI in 2013 to be a set of conditions rather than a specific location. Those conditions include proximity of structures to vegetation, weather patterns, fire history, topography and access, among others. The NFPA's reliance on conditions creates a far more flexible definition capable of embracing different WUI conditions nationally as well as locally. The International Code Council uses a similarly flexible definition in its Wildland-Urban Interface Code: a geographical area where structures and other human development meets or intermingles with wildland or vegetative fuels.

The Human Factor

As an environmental historian by training, I've started thinking about the WUI from different perspectives in the hope of inspiring meaningful risk reduction efforts. I see landscapes and ecosystems as historical and dynamic. Ecosystems are complex. As historian Mark Fiege explained convincingly, ecosystems are patterns and products of "reciprocal interactions between human systems and water, landforms, soil, plants, animals and climate." Modifying one aspect of an ecosystem, such as replacing growing fuel with built fuel, has short-term and long-term consequences. It's impossible to erase the flow of time and the impact of time on an ecosystem, landscape, nation-state or person, for that matter.

Humans are one more natural component of a natural system.

Ongoing misguided political, cultural and philosophical attempts to divorce humans from our ecosystems only make matters worse. Aldo Leopold captured this idea in his seminal *Sand County Almanac*:

> *That man is, in fact, only a member of a biotic team is shown by an ecological interpretation of history. Many historical events, hitherto explained solely in terms of human enterprise, were actually biotic interactions between people and land. The characteristics of the land determined the facts quite as potently as the characteristics of the men who lived on it.*

In addition to modifying our relationship with fire, we won't be able to make a solid difference in our collective safety until we recognize and appreciate that *Homo sapiens* is one more critter that must follow the same rules as all the other critters in the biosphere.

Why modern Americans live in fire prone ecosystems and why they build homes to mimic a previous or imagined ecosystem is considered by environmental historian Elizabeth Carney. She also captures why new residents landscape the excavated dirt with little respect for combustible host ecosystems. According to her research, this aspect of the WUI story and our relationship with wildfire can be traced back to the 1930s when middle-class Americans acquired the means to move away from cities and began to do so in earnest.

Sprawl has become a profanity in the American lexicon. I learned to use it in the 1980s while riding in the backseat of my parents' Dodge Dart between our home in Idaho Springs and the westward expanding neighborhoods and stores of Wheat Ridge where my maternal grandparents lived. Sprawl also described the southward growth of the Denver metropolitan area toward my uncle's home in the ranchette-strewn hinterlands of Douglas County known as Surrey Ridge and Lone Tree, which are now solidly within the suburban service area of South Metro Fire Rescue Authority. Yet sprawl isn't a fair description of why developers built those homes along the ever-moving perimeters of towns and cities or why they built homes either in pods or individually along dirt roads in valleys, on mesas, along ridges, or in prairies previously uninhabited except by culturally diverse nomadic and itinerant hunter-gatherers or resource prospectors.

That growth along the metropolitan perimeter and in rural areas is explained in part by a cultural paradigm shift that favored outdoor-living, a philosophy that commodified ecosystems into tidy, sanitized packages. These newly developed ecosystems had everything new residents desired: urban opportunities, natural amenities and wilderness access "where Rocky Mountain penstemons, California redwood decks and exotic trees lived together in affluent harmony." As people fled the no-longer-coveted urban centers and gladly bought the outdoor-living package, they encountered long-term residents of those same ecosystems: ranchers, farmers, miners, loggers, and dwellers of extraction-based towns.

It wasn't enough to move to the mountains and their valleys. Developers fashioned homes to resemble the idyllic ecosystems surrounding their abodes. They used wood siding, hand-split cedar and pine shake shingles for roofing, wood decks built around trees and both native and non-native vegetation planted beside the house to camouflage it. Where developers left off, architectural committees and homeowner associations crafted covenants to maintain the "natural" aesthetic standards. Few of those standards and designs recognized the wildfires that created—and someday would return to—the host ecosystems.

Open spaces and opportunities to build homes and lifestyles within them made the American West exceptional, but the natural look of its architecture was exported through suburbs and back to city centers. Suburban homes in neighborhoods named for prior ecosystems such as Pine Valley, The Timbers, Oak Hills, Willow Creek, and Castle Pines looked like their surroundings. Unbeknownst to many of their owners, the houses also burned like the native vegetation they represented and replaced. In fact, this replication of native ecosystems and their processes of change is an ongoing legacy of the outdoor-living aesthetic period.

More People Need More Homes

Even with plenty of examples to suggest such growth creates more costs than benefits when wildfires do return to an ecosystem, it continued throughout the latter half of the 20th century and hasn't changed in the 21st.

While living in nature is a motive, population growth also drives development in wildfire-prone ecosystems. The population of western states tripled between 1950 and 2000, while the US population as a whole didn't even double, according to the U.S. Census. The population of the West grew faster than the Midwest, Northeast or South in each decade of the 20th century. The West's population represented only five percent of the nation's population in 1900, but now is close to 25 percent.

More residents require more living units. We can build vertically or horizontally. As the Denver skyline can attest, creating architectural masterpieces that seem to scrape ever-deeper into our skies is one desired option, but many other consumers prefer to build, purchase and inhabit homes away from urban centers, pushing human constructs—concrete and abstract—into "the woods." Land availability makes that growth possible. In 2000, the population density in the Northeast was 330 people per square mile while in the rapidly growing West the density was only 36 people per square mile.

We need to follow the money to understand our WUI situation more completely. People are allowed to live in wildfire-prone ecosystems because they are willing to buy those homes, but I've wondered for years why county officials—elected or hired—seem to ignore public safety when they permit home construction. I've served in the fire service in three partially rural Colorado counties and found private properties in each that left me shaking my head why various county decision-makers allowed such designs to be constructed.

Narrow, steep driveways with switchbacks usually catch my eye initially because firefighters can't help residents if they can't access residents. Homes perched on steep mountainsides with wood roofing and surrounded by dense native and ornamental vegetation confound me. While firefighters, life safety educators, land managers, and advocates from non-profits flood media with messaging about mitigation, county leaders seem to take a more liberal interpretation of prevention and mitigation, ultimately siding with the generation of tax revenue and apparently shirking their part of the burden when wildfires burn homes.

Colorado's voters set the stage for such growth in 1982 when they passed the Gallagher amendment to the state constitution. This amendment lowers the assessment rate—currently under eight percent—on residential property when residential property values grow faster than nonresidential property values. Yet the amendment maintains the nonresidential property assessment rate at 29 percent. Commercial properties get the media attention, but vacant land including land categorized as agricultural also gets taxed at this high rate. Arguably, building a house or cabin on your 35-acre parcel is economically more responsible over the long term than letting the land sit vacant because the tax rate plummets from 29 to less than 8 percent. That trend will continue as long as local and state governments can reap the benefits of growth and look to federal budgets to cover the costs of fire suppression. Until then, they or their constituents have little incentive to restrict or regulate growth in wildfire-prone ecosystems or place responsibility on property owners.

The proof is in the pudding. The number of homes built in high or very high risk wildland urban interface settings is already stunning. Colorado, Montana, New Mexico and Oregon have the highest percentages of homes in those top risk categories, according to a report published by the Union of Concerned Scientists. The states with the greatest number of homes in those risk categories are Texas (678,000), California (375,000) and Colorado (200,000).

As described in a previous chapter, risk perception is integral to this discussion. When my family lived in a wildfire-prone ecosystem, we did so consciously. My wife and I were both firefighters who understood the risk that our growing and built ecosystem posed: plenty of fuel produced by a fire-friendly ecosystem, a neighborhood protected by a small cadre of volunteer firefighters at the end of a single access road ten long miles from the next nearest fire department resources, a limited water supply. But recreation, security, privacy, aesthetics, a tight-knit fire department and wildlife—from snowshoe hares to black bears on our property—outweighed the risk of wildfire.

Mitigation's Return of Investment

The wildland urban interface will continue as part of our nation's narrative indefinitely, thus finding a new relationship with fire is essential. There have been pioneers in this relationship building. The story of Alpine, Arizona, appears above. Another success story is the Rancho Santa Fe Fire District in San Diego County, California.

Residents of five communities within that fire district—The Bridges,

The Crosby, Cielo, 4S Ranch, and The Lakes—live in subdivisions designed for wildfire-prone ecosystems. As the fire district's website explains:

> *Every home in your community, including yours, was built to certain standards that, when maintained, greatly increase the chances that your home will survive a wildfire. In fact, while we encourage everyone to evacuate early if a wildfire approaches, it may be safer to stay in your home, or shelter-in-place, than to evacuate under hazardous conditions.*

Those homes are constructed of fire-resistant materials and include characteristics such as residential sprinkler systems, boxed eaves, non-combustible roofing, dual pane or tempered glass windows, and fire-resistant landscaping covering at least a 100-foot radius around the structure. The community also meets certain standards including roads wide enough for two-way traffic consisting of large firefighting apparatus, adequate water supplies, and defensible space around the community itself. All of those characteristics are maintained through-out the year and inspected for compliance.

Most civilian wildfire-related deaths occur during evacuation efforts when heavy smoke and ember blizzards obscure roadways congested with freaking-out drivers. Rather than prevent wildfires, residents, planners and firefighters in San Diego County prevent the need for evacuation. Rancho Santa Fe Fire Chief Erwin Willis established the shelter-in-place program in 1997. Ten years later, after countless debates, it proved its value when the Witch Creek Fire—part of the California Fire Siege—raced through San Diego County in October 2007. That fire burned nearly 198,000 acres and destroyed 1,624 homes. When it approached the shelter-in-place communities, the fire burned around them rather than through them. No homes were lost in those communities. Building communities to resist fire is an idea long overdue in the United States.

A pair of fires in Colorado also highlight the potential return for the investment of mitigation although they are better known for their failures than successes. Driven by 65 mph winds, the Waldo Canyon Fire exploded from the forest into the northern suburbs of Colorado

Springs on June 26, 2012. Over 30,000 residents had evacuated in advance of the rapidly expanding firestorm. Home destruction ended as a third round of winds diminished and exhausted firefighters halted the ember-initiated home-to-home urban conflagration. Two people died and 347 homes burned to the ground. Yet residents of one neighborhood in the fire's early path who had invested in mitigation—those living in Cedar Heights—escaped scared but unscathed.

Residents had invested in mitigation, but so had the Colorado Springs Fire Department. Its wildfire mitigation team used pre-disaster mitigation grants from FEMA to assess wildfire risk throughout the city, conduct parcel-level risk assessments, and create mitigation projects in coordination with other city agencies and homeowners. Mitigation crews thinned open space around Cedar Heights and chipped slash generated by residents of the neighborhood. Investing roughly $300,000 into mitigation prevented $75 million in losses, had the neighborhood's 250 homes and water tanks burned.

The CSFD also invested time in teaching city residents about mitigation and for evacuations. As one resident of the Mountain Shadows neighborhood later shared, "I learned how to identify items in my house that I would plan to take, and those I would leave, in an evacuation situation." That training paid its dividends. On June 24, 2012, while returning from the grocery store, "We could see smoke and flames. We unloaded the groceries and 20 minutes after getting home, we received a 911 reverse call to evacuate immediately." Two days later, "we were at the 4 p.m. daily press conference when the fire came over the hills into Mountain Shadows. I remember the mayor asking everyone to evacuate immediately, in the middle of the live conference. We drove back to our hotel and watched the fire destroy our neighborhood into the evening." Her house survived the conflagration, but the trauma of those losses remains.

One year later, a fire ripped through the Black Forest community northeast of Colorado Springs burning 511 homes and 14,000 acres. Mitigation worked for many homeowners, but not for all of them. Said one resident, "My home was mitigated but it burned, along with the rest of the neighborhood, in a raging, extremely hot frontal attack crown fire. Had it been a surface fire with milder conditions, I believe my mitigation would have worked well."

A Success Story

Individual mitigation wasn't enough in that blaze, but collective efforts—as occurred in the Cathedral Pines neighborhood of Black Forest—was effective.

The eleventh day of June 2013 was another red flag day for northern El Paso County, Colorado, according to the National Weather Service. Meteorologists predicted yet another day of sustained southwesterly winds between 15 and 25 mph with gusts to 40 mph and relative humidity as low as four percent. Combine that with dry fuels, temperatures in the 90s and 100s, and a Haines Index rating of 6 (which indicates the lower atmosphere was unstable) and there was a high potential for fire growth.

In a stable atmosphere, hot air rising above a fire spreads horizontally because it is cooler or the same temperature as the surrounding air. In an unstable atmosphere, the hot air is significantly hotter than all of the air above it and it can ascend unchecked to the upper atmosphere in a defined column, generating wind and lightning as it grows.

The red flag warning stated: "extreme fire behavior is imminent if a fire starts." Conditions were ripe for a wildfire, especially along Colorado's Interstate 25 corridor. They were so ripe that a few minutes before noon Black Forest Fire Rescue Assistant Chief James Rebitski requested firefighters to staff the agency's stations voluntarily to augment paid crews already on duty.

Less than two hours later, Rebitski's hunch had delivered. Black Forest Chief Bob Harvey noticed a smoke column southwest of his station but thought it might be related to a large wildfire growing near Canon City over 60 miles away; civilians began calling 911 at 1:42 p.m. to report a smoke column northeast of Colorado Springs. Without a specific location, several fire departments were dispatched to find the fire as soon as possible. Firefighters began converging on the fire south of Cathedral Pines at 1:50 p.m. They found medium to heavy smoke but no flames. The wind was out of the southwest at 29 mph, the temperature was 95 degrees, and relative humidity was at four percent, according to the Black Forest Fire's Investigative Report.

One of the first firefighters to reach the fire estimated its size as 50 feet by 50 feet, but he couldn't see actual flames through dense vegetation growing between homes. He and others began structural protection activities as the fire expanded in size and intensity. Initially, flames were two to four feet in length, which meant firefighters could engage the fire directly, but at 2:18 p.m. the fire blew up. It began igniting multiple homes, showering the area with a blizzard of embers and crowning: burning rapidly through tree canopies with flames lengths much too great for firefighters to engage safely. The crown fire sprinted north and east, pushed by the strong winds and fed by plenty of growing and built fuels.

Cathedral Pines, a subdivision of 175 homes on lots between two and five acres in size, was due north of the fire's origin. Development began in the 1990s, but followed a philosophy ahead of its time. Mitigation accompanied home construction. The developer thinned trees and brush, cut low-hanging branches (called ladder fuels because they enable flames to climb into canopies) and paved wide roadways, according to a report from the Pikes Peak Wildfire Prevention Partners. Additionally, covenants were written to require fire-resistant building materials and fire-resistant landscaping. Residential sprinkler systems were required in homes greater than 6,000 square feet unless homeowners installed hydrant systems or cisterns in the neighborhood, or contributed to a rural fire fund for the county.

When the crown fire reached Cathedral Pines, it dropped to the ground because these tree canopies weren't interwoven as they were in other parts of the Black Forest community. The fire burned along the surface fuels—grasses, dead pine needles—and released enough convective heat to scorch some lower branches, but few trees ignited from that process. Embers showered the neighborhood while the surface fire advanced around homes, but those embers ignited and subsequently destroyed only a single house in the mitigated neighborhood. The wide roads interrupted the spread of the surface fires. Ultimately, the neighborhood was built and maintained in such a way that firefighters could operate safely and thus actively and successfully defend 175 homes from the wildfire.

Firefighters racing to defend homes throughout the rest of Black Forest weren't as fortunate. The fire burned 14,280 acres, killed two people still in their home, destroyed 486 homes, and forced over 38,000 residents to evacuate. The El Paso County Sheriff's Office estimated the total market value loss at $116 million. Indirect losses continue.

Restoration Is Misguided

Much of the existing risk reduction efforts in WUI messaging espoused by foresters, firefighters, politicians, and life safety educators is based on the myth that our forests are denser than they would be "naturally." The idea that the pre-suburban West was natural and the current ecosystems are unnatural or artificial is myopic and frustrating. Myths, as historian Mark Fiege explains, "reduce the complexity, ambiguity and uncertainty of human experience to simple, timeless, viscerally compelling stories." Myths, like models, are useful for testing theories, but they should be used with caution while remembering that they are simplifications of something greater. As statistician George Box said, all models are wrong but some are useful.

A photo of a bride and groom, for example, is a model of a wedding. While it captures the joy and wonder felt by the newly married couple at that exact moment, it doesn't capture the romantic wooing that inspired the vows, the logistics of organizing the ceremony, the awkward speech by the Best Man, Great Aunt Edna snoring during the service, or the first dance with the in-laws. Models are limited by the amount of detail and context they can replicate. They are useful, but only when used with a critical mind that recognizes their limits and the observer's bias.

The assertion that human relations with the Earth, and particularly modern Americans' relations, are always destructive, as Aldo Leopold stated, is a "fundamental conceptual barrier" to reducing risk in WUI areas. That philosophy drives restoration advocates who believe that a century of fire suppression and four centuries of American growth have interfered with the natural order.

Restoration—the belief that society needs to manage its forests to match or re-create a natural order based on the number of trunks in an image captured by a photographer in the 19th century or based on a computer model—is flawed. It has as much traction in the political and practical realms of wildland urban interface and forestry circles as the natural order myth. The wildland urban interface is neither a restored nor a destroyed pristine nature. Nor is it an unbalanced nature. Each of these oversimplifications ignores the historical essence of our planet, the complexity of ecosystems and the human role in our ecosystems.

Restoring an ecosystem to a "natural" state sounds appropriate and scientific, but intuitively the idea leaves a bad taste on the palate. As James Malin said regarding the history of the grasslands: "Any attempt at regional definition of a portion of earth space involves time as one of the determining criteria."

Judging a current ecosystem as a degradation of its former self (or improvement, for that matter) based on a snapshot from a photographer or painter or diarist from a century ago denies the role of time, observer's bias, and the whole concept of history. As much as one of William Henry Jackson's images may differ significantly from an image snapped today, his view might have differed significantly from that view 100, 50 or even a handful of years before he tromped through those woods. It also could vary if he adjusted the aperture to change depth of field. The images hide the ongoing roles of climate, weather, humans, other animals, and myriad other factors.

Animals, climate fluctuations as identified by dendrochronologists (tree ring folk) and palynologists (pollen people), and dynamic weather patterns play roles, especially in transition areas that hinge life zones together. Climate also influences the toll insects, disease, humans, wildlife, and wildfires take on vegetation.

Malin rightly argued that all those factors—not the least of which is time—"raise an honest doubt as to whether the idea of climax vegetation is even legitimate or of practical value." Indeed. Climax or any restoration goal is static and linear. It declares a pinnacle of achievement and thus oversimplifies a complex system. As former USFS Chief Jack Ward Thomas is credited with stating: "The ecosystem is not only more complex than we think, it is more complex than we can think."

In 1897 the National Forestry Committee suggested that the nation's forests be preserved with "systematic and intelligent forest reproduction" strategies to protect water supplies and the supply of forest products demanded by the industrializing country: "The enormous waste from forest fires, incendiary and accidental, which prevail in nearly every part of the United States...threatens the prosperity of the country."

Government regulations, developed from the bias of progressive industrialists, also influence forest growth. Government theories about managing nonhuman factors as if they were components of a machine have modified ancient fire patterns and understandings at least as much as home construction and suppression efforts.

Restoration goals underestimate the complex fluctuations and relationships in that place. They rely on human values and assumptions about those values. Is a forest with a given number of stems the best status of a given acre? Foresters and loggers might think so. Why is a forest better than a grassland? Ranchers might disagree. Deer, voles and sparrows might prefer growth of Gambel oak, mountain mahogany and skunk brush. Maybe homes on three-acre lots liberally sprinkled with pampas grass, junipers and Austrian pines are the best use. Developers, realtors and future residents might vote for this option. Whose values are most valuable?

Asking whether Colorado's Front Range forests are "overgrown" is an academic exercise akin to estimating the number of angels on a pin. Asking whether that growth is a product of decades of fire suppression is valid, but lacking a control group the answer is elusive. The orthodox answer is yes. A historically sound answer is maybe. Ecosystems are more complex than we can think.

Restoration as a status is misguided, but as a process in which fire is allowed to burn through ecosystems it makes sense. I strive to convince our residents to harden their homes against wildfire so that when wildfires do return, they weave through subdivisions recycling vegetation while leaving structures unscathed. Historically speaking, restoring fire to fire-adapted ecosystems seems more responsible and possible because wildfire is part of the historic harmony of that ecosystem. However, new fuels in the form of homes and the presence of people make this strategy complex, too. Managing a parcel of land based on a black-and-white photograph, preventing fires outright, deconstructing subdivisions or prohibiting future development are counter-productive, expensive and ineffective.

Finding harmony is effective. For me it represents resiliency, balance, flexibility and strength. My interpretation of harmony, a relationship that evolves with time, guides my understanding of my surroundings and my risk-reduction efforts.

Finding Harmony

As a new environmental historian back in the day, I embraced the concept of a hybrid landscape, in which reciprocal interactions of human and non-human components of an ecosystem create a nebulous, dynamic nature. As humans respond to the characteristics or environmental parameters of a given ecosystem, they modify the ecosystem and the parameters of future interactions. I also explored the idea of equilibria in ecosystems, but too many interpretations divorced humans from our surroundings or suggested Nature was either in balance or out of balance.

My own research and observations of our ecological context denies that nature functions linearly or in a binary either/or format. So I turned to my Dad's field of physics. In a 1950s college text book, physicists related equilibria to a cone:

> *A right circular cone on a level surface affords an example of the three types of equilibria. When the cone rests on its base, the equilibrium is stable. When balanced on its apex, the equilibrium is unstable. When resting on its side, the equilibrium is neutral.*

The threat to equilibrium is a disturbance or stress. As Le Chatelier's Principle describes, "If a stress is applied to a system at equilibrium, then the system readjusts, if possible, to reduce [the] stress." Systems in stable and unstable equilibria are unable to readjust. When the cone tips over, it's done. However, a system in neutral equilibrium can adjust. It won't be the same—we can't undo the disturbance or reverse time—but as the cone rolls on its side, a system remains functional.

When a wildfire burns through a forest of lodgepoles, some trees and surface vegetation die while the ecosystem lives. Lodgepole seeds spring from serotinous pinecones and germinate in the soil, grasses grow in new patches of sunlight and in time lodgepoles may return to their former height and density. In the meantime, the mountain pine beetle population declines because its preferred food and breeding area are gone. Beetles move elsewhere. Other animals also change their habits in response to the wildfire and its impacts on the vegetation, soil and snowpack.

The cones of equilibria are useful, but they are too simple. That hunger for complexity led me to the harmony of the Navajos: hózhǫ. I only have a shallow understanding of hózhǫ thus far, but it explains the responsibility humans have in our dynamic ecological contexts. My first encounter with this intriguing idea came from the crime novels of

Tony Hillerman. A crime novel may not be the best resource for cultural understanding, but it is a good starting point. A brief explanation from *Sacred Clowns* sheds some light on this vision of harmony:

> *Terrible drought, crops dead, sheep dying. Spring dried out.*
> *No water. The Hopi, or the Christian, maybe the Moslem, they*
> *pray for rain. The Navajo has the proper ceremony done to*
> *restore himself to harmony with the drought. The system is*
> *designed to recognize what's beyond human power to change,*
> *and then to change the human's attitude to be content with the*
> *inevitable.*

That sense of harmony is satisfying and recalls a comment from Leopold about changing "the role of *Homo sapiens* from conqueror of the land-community to plain member and citizen of it." Change our attitude to accept wildfires as inevitable and empower us to modify our homes and neighborhoods in advance of the next wildfire. Such a strong, resilient relationship enables responsible community risk reduction.

Believing that ecosystems cease being natural because we humans live in them has disastrous implications for both humans and our nonhuman neighbors. Forest restoration suggests that if ecosystems are constructed, developed and otherwise meddled with to resemble a snapshot from the past, that all will be right with the world.

However, maintaining the natural setting or restoring it to its appearance during the pre-house era will not stop WUI fires, especially the catastrophic mega-fires that incinerate property and kill people. Predicting the future is difficult anyway, but predicting the future based on incomplete ideas about the past and faulty impressions of the present is foolhardy. "Each space-time situation is the product of a unique combination of factors which can never be brought together again," Malin rightly argued. Any changes to an ecosystem have short- and long-term consequences that are irreversible.

Every act happens only once. We need to accept the historical essence of our wildland urban interface and our impact within it if we have any hope of living here safely. We need to change our attitudes to accept that fires are possible and, from a greater perspective, inevitable.

The time for hózhǫ is upon us. ⌒

10

Will We Keep Burning?

...man in his originals seems to be a thing unarmed and naked, and unable to help itself, as needing the aid of many things; therefore Prometheus made haste to find out fire, which suppeditates [sic] and yields comfort and help in a manner to all human wants and necessities; so that if the soul be the form of forms, and if the hand be the instrument of instruments, fire deserves well to be called the succour of succours, or the help of helps, which infinite ways affords aid and assistance to all labours and mechanical arts, and to the sciences themselves.

Sir Francis Bacon

Listen, if you want to live out in the woods, that's great, but you shouldn't put a wood-shake roof [on your house].

Colorado Governor John Hickenlooper, 2013

*A*s ancient and modern stories reveal, fire is sacred to us, both spiritually and secularly.

The following story of Goorda from the Aboriginal people of Australia demonstrates that fire is fragile and powerful, capable of cooking meat and incinerating landscapes. Fire is complex rather than easy, and requires a set of rules to use harmoniously:

> *Goorda, the fire spirit, lived alone among the stars of the Southern Cross. He was a great hunter who traveled between three different campfires, known as the Pointers. When he was lonely, Goorda wanted his neighbors to visit...*
> *"I come bearing a gift for you," said Goorda. When his feet*

touched ground, however, the brush began to crackle with flames. Soon, a blaze was roaring along the riverbank. Across the river, the people picked up their spears and watched, their faces kissed by the red hues of the dancing flames.

When the heat from the fire became so great they could feel its pain, they screamed and fled from Goorda.

"Stay! Do not go!" Goorda cried, "I will not hurt you." But flames spread everywhere he walked. Those people who jumped into their canoes and paddled away from shore survived...

But Goorda, not understanding, tried to get close to people so he could visit with them. Some people were burned in their huts.

Others fled in terror. By the end of the day, Goorda looked out over a silent, black, smoldering land. Overhead, a red-winged parrot called and circled in search of a place to land, then flew away searching for safety until it sank down below the horizon in the distance.

Exhausted, Goorda sat down to rest. He found a kangaroo that had been burned and began to eat. "This will not do," he said.

"I will have to appear to the Earth people in a different way, or they will always flee from me." Goorda changed to the form of a person and painted a diamond on his breast. "This will be my symbol," he said. "Now I am ready to greet people and teach them how to use fire in a good way."

When the sun rose over the blackened land Goorda had created, a curious group of hunters came wandering. Goorda saw them and stood up. "Hello, my friends," he said. Flames began to shoot from the end of the stick he held in his hand. "Do not fear," said Goorda. "Fire can be of use to you. It can cook your food." Saying that, Goorda picked up a piece of blackened kangaroo, took a bite and chewed.

"See, it tastes much better this way."

Goorda handed the meat to the men and each carefully took a bite as they passed it around. "Indeed, it is very good," they agreed. They searched around for more meat and ate their fill. As they were eating, Goorda showed them the ways of fire. He demonstrated how to create fire by holding a stick between the hands and turning it quickly while the end sits in a small hole *bored into another piece of wood. In a short time, the light of an ember appeared where the two sticks met.*

"Be careful with flame," said Goorda. "Clear the brush away from your campfires and watch the flames closely so they do not spread.

"When the leaves and branches are dry, you can light a fire to herd the animals so hunting will be easier. Later, when the rains come and the Moon Man has once gone and returned, those trees and bushes will sprout new leaves and twigs for the wallaby and kangaroo to eat."

The hunters learned everything Goorda had to teach, then they thanked him and walked back to tell their families...

In order to build hózhọ, we must be careful with fires. Although recorded fire prevention in this country only dates to the colonial era, Native American cultures had their own regulations governing prevention, such as reducing fuels around their communities prior to burning in order to influence hunting and harvesting patterns. Colonial Americans did the same. In 1648 Dutch colonists instituted a fire prevention ordinance in New Amsterdam (New York City) regulating building materials and mandating frequent chimney cleanings and maintenance. Fire wards were charged with watching for fires, coordinating bucket brigades and enforcing building codes in Boston's neighborhoods in 1712 after decades of blazes at regular intervals leveled blocks of the city. Other cities across the growing nation invariably adopted more fire-resistant building codes and advocated safer practices as a result of fires that destroyed property and killed citizens.

Fire prevention became institutionalized in our American culture on October 9, 1911, by the Fire Marshals Association of North America (FMANA), 40 years after the most destructive days of the Great Chicago Fire and Peshtigo wildland fire. As described earlier, Chicago's firestorm killed at least 250 people, destroyed over 17,400 structures and left 100,000 citizens homeless. Simultaneously a couple hundred miles to the north, the Peshtigo fire, which would now be known as a wildland urban interface fire, razed 16 towns, killed more than 1,100 people and blackened 1.2 million acres. October 1911 also was six months after the deadly Triangle Shirtwaist Factory fire in New

York City and 14 months after the storm of wildfires in Washington, Idaho, and Montana now known as the Big Blowup. Members of FMANA decided that the anniversary of those fires should henceforth be observed not with festivities, but in a way that would keep the public informed about the importance of fire prevention. Thus, they established Fire Prevention Day.

Approximately 200 of the nation's experts in the fire service, industry, media, and government gathered in Philadelphia for the First American National Fire Prevention Convention in October 1913. Hosted by the Fire Waste Committee of the U.S. Chamber of Commerce, they considered a variety of topics to address the heavy burden fire was placing on their nation in terms of deaths and property loss in order to initiate a movement to educate the public about fire danger and waste. They also sought to study, prepare, enact, and continuously enforce appropriate minimum legal requirements to protect life and property from fire throughout the country.

In step with the Progressive Era philosophy of scientific management, attendees largely agreed that individuals were granted too much liberty to decide on their own how to construct, equip, occupy, and manage property. "Greed, ignorance, indifference and shiftlessness" usually replaced any concerns for safety and thus permitted "a frightful loss in life and property."

Specifically addressing educational matters, the assembled throng agreed with Franklin Wentworth, secretary of the National Fire Protection Association (NFPA), that each state should have a state fire marshal to serve as educator and prosecutor of arson cases and each fire department should conduct inspections of properties. They passed a resolution:

> *Education of the public about fire danger and waste of life and property should be provided in all laws, ordinances and regulations on the subject; and all interests concerned should not only join issue in collecting accurate and authoritative data but make equal effort to disseminate this information regularly and continuously among all people in readily understandable language, to the end that they may not only accept but demand proper fire waste regulation and live in full accord therewith.*

Wentworth and the NFPA took that recommendation to heart, adding educational efforts to its preexisting agenda to pass improved standards and codes.

President Woodrow Wilson issued the first National Fire Prevention Day proclamation in 1920. In 1925, President Calvin Coolidge expanded the commemoration to the entire week surrounding October 9, the anniversary of the worst day for both Peshtigo and Chicago. Additionally, over 15,000 people had died in fires during his first full year in office, prompting President Coolidge to remark:

> *This waste results from conditions that justify a sense of shame and horror; for the greater part of it could and ought to be prevented. It is highly desirable that every effort be made to reform the conditions that have made possible so vast a destruction of the national wealth.*

President Harry Truman called for and participated in the President's Conference on Fire Prevention in May 1947. In the years preceding the event, as many as 10,000 people had died annually in fires or from burns resulting from fires, and wildfires had consumed several million acres of forests. Said President Truman:

> *I can think of no more fitting memorial to those who died needlessly this year in the LaSalle Hotel fire in Chicago, the appalling disaster at the Winecoff Hotel in Atlanta, and the more recent New York tenement holocaust than that we should dedicate ourselves anew to ceaseless war upon the fire menace.*

Among the achievements of that often-forgotten symposium was the identification of the "3-Es" by the participants. Education, Engineering and Enforcement were and remain integral to preventing fires as well as injuries, illnesses and property loss. Engineering can make the built-environment safer. Enforcement adds positive and negative consequences—generally fiscal—to help influence people to change their behaviors and environments. Neither is effective without education. The goal of education is to change ideas and attitudes that result in changed behavior. Without education, engineered solutions are ignored or misused and enforced standards are forgotten when an inspector walks away.

Despite national and local attention to fire prevention, the number of ignitions, injuries and deaths increased. President Richard Nixon added impetus to the need for prevention as part of the National Commission on Fire Prevention and Control's 1972 report *America Burning.* "Only people can prevent fires," President Nixon wrote.

> *We must become constantly alert to the threat of fires to ourselves, our children, and our homes. Fire is almost always the result of human carelessness. Each one of us must become aware—not for a single time, but for all the year—of what he or she can do to prevent fires.*

Only one of the report's conclusions about decreasing the annual toll of fire concerned suppression: firefighters needed better training and education. The other five conclusions and the bulk of the report covered prevention and mitigation topics, emphasizing fire prevention within the fire service, educating Americans about fire safety, eliminating unnecessary hazards from the nation's built environment, improving fire protection features such as sprinkler systems and smoke alarms, and expanding research into fire behavior in order to prevent it, slow its growth and find new ways to detect its presence.

Four decades have passed since the release of *America Burning.* Some progress has been achieved, but fires keep burning, injuring, and killing in our nation. Since the 1970s, the number of home fires throughout the United States has decreased, but your chances of dying or being injured in a fire have barely improved.

Long forgotten as a rite with rules, fire itself is now viewed as a right—nearly constitutional in stature—and linked with property rights, privacy, choice, and freedom from government regulation. Fire bans, fireworks bans, rules about candles, covenants regarding open burning and similar restrictions meet staunch resistance until a fire damages property, injures or kills. At that point, governments, homeowner associations and landlords are blamed for not having policies in place to prevent fire's negative consequences.

Colorado Governor John Hickenlooper was interviewed on Public

Radio following a tour of a neighborhood scorched by the 2013 Black Forest Fire. Host Ryan Warner asked the Democratic Governor if the state should require homeowners to mitigate around their properties or require some form of a unified statewide building code. Hickenlooper's answer revealed his own philosophies, of course, but he also seemed to palpate the pulse of residents:

> *I spoke last week to the Colorado Municipal League, which represents municipalities and communities all over the state; up in the mountains and out on the plains. We discussed it. We don't do a statewide building code. We're a state that is focused on local ordinances. I laid it out there in pretty blunt terms saying, 'Listen, if you want to live out in the woods, that's great, but you shouldn't put a wood-shake roof [on your house].'*
>
> *Our job is to make sure the local officials get the facts and how devastating it is if you don't tighten up on these building codes. People want us to be partners with them and help them define what exactly defensible space looks like and how do we minimize forest fires as much as humanly possible...*
>
> *...but that doesn't mean the state should be dictating or usurping the local role.*

He's right. Government mandates can't replace individual responsibility. Personal responsibility gets lost in post-fire blame games, but it is among the most powerful safeguards when we interact with fire. One of my friends, Lisa Miklas, has embraced responsibility as a defense measure since her family's business burned in the 1970s:

> *When I was young, my family owned and operated a live-stock auction in Vancouver, WA. My mom was getting me ready for my first day of kindergarten when we got a phone call that there was a fire at the auction. As we got near, we could see, and smell, the smoke.*
>
> *We were very fortunate that it was a Tuesday, and there were few animals or customers on site. There was chaos everywhere. The truck of one of our employees was in the parking lot, and it was not known if he was inside the business or not.*

One of our barn cats had kittens on site and was anxiously try-ing to get them all out. It was all very overwhelming for me at age 5. After we realized that our employee was safe and sound elsewhere, my grandmother took me home and left my mom at the auction with some of our employees and friends.

The experience was nearly forty years ago, but has left an impact on me for life. There are still times when I wake up in the night with real concerns about fire safety and cannot get back to sleep. I realize that a fire can happen to anyone, anywhere; it is not just an abstract idea. The reliability of working smoke and carbon monoxide detectors in my home is non-negotiable.

As Lisa explains, we as individuals must share the responsibility for reducing the risks of all fires to our families and properties.

International Strategies for Safety

A willingness to share responsibility for fire safety is one reason other nations have lower accidental fire death rates than the United States. The National Center for Injury Prevention and Control conducted a study of other nations' fire protection efforts in 2003 and 2004 to deter-mine if their best practices might find similar success in our nation. Those targeted countries—Australia, New Zealand, Japan, England, Scotland, Norway and Sweden—have created myriad programs and standards to guide their efforts:

- Prevention programs developed at state or national levels;
- Fire safety programs that enhance state and national curricula for schools;
- Workshops for caregivers of high-risk audiences such as preschool-ers and elderly citizens;
- Regular program evaluation that often results in programs getting shelved or at least redesigned;
- Firefighters visit children who set fires (with their parents) in their own homes;
- National building codes set standards for homes in wildfire-prone ecosystems;
- More personnel trained for prevention roles;

- Push responsibility for fire inspections onto building owners; fire-fighters are secondary partners for safety.

British firefighters have reduced their fire death rate by 40 percent since 1990, in part, because they now visit high-risk homes to verify the presence of working smoke alarms. Could such a national program be imported to the US? Cultural, political and economic differences present some obstacles to adopting similar programs in America, but change can be a good thing if we decide collectively that fire safety is a worthwhile investment.

Australia's "leave early or stay and defend" policy is based on sharing responsibility during wildfires. Residents are encouraged to harden their homes and properties to resist wildfires and to leave early—before road conditions deteriorate—if a wildfire threatens their community. Residents who don't leave early are trained (in advance of a fire) to stay and defend their properties. Australia's fire service has reduced the number of last-minute evacuees and provided additional firefighters able to knock down small fires to save homes.

The policy was scrutinized unfairly in 2009 following the fatal Black Saturday Fires. On February 7, residents of the state of Victoria were warned about catastrophic conditions: a record heat wave was baking the state and winds above 50 miles per hour were strafing the landscape. Dozens of wildfires ignited including the Kilmore East Fire after a power line broke or sparked over dry vegetation. That small fire exploded into an inferno that raced through small communities, overwhelming firefighters and surprising residents. Had many of those residents evacuated earlier or stayed in their hardened homes, many more could have survived Black Saturday. Instead 173 people died that day.

One reason fire can be wielded with little thought about its potentially devastating consequences in this country is the strength of the American fire service. Firefighters, life safety educators, fire inspectors, chief officers, mechanics, dispatchers, and the administrative staff that keeps all of them functional, whether paid or volunteer, are serving almost every inch of our nation and its people. When men and women are willing to train, wear special clothing and bring special equipment into a fire, it's easy to forget that fire and its byproducts are killers.

Because firefighters are among the dead each year, the American fire service has adopted 16 Life Safety Initiatives to help everyone go home safely at the end of the incident, shift, and career. Unlike the content of *America Burning*, the content of the 16 Initiatives leans heavily toward reacting to emergencies. Only two of them explicitly address prevention:

14. Public education must receive more resources and be championed as a critical fire and life safety program.
15. Advocacy must be strengthened for the enforcement of codes and the installation of home fire sprinklers.

I'm fortunate to work for an agency that commits resources to community risk reduction, but my agency is among a minority.

Smoke Alarms

Note that smoke alarms aren't emphasized in those initiatives. They now are accepted and revered as an accessible life safety technology, but they're not ubiquitous. Research suggests they are missing from four percent of U.S. homes, but they are prevalent enough that they didn't warrant specific mention in Initiative 15. Smoke alarms are installed to give occupants more time to escape and, in monitored systems, to notify firefighters as quickly as possible. They weren't automatically welcomed into homes, though.

Francis Upton and Fernando Dibble are often mistakenly credited with inventing smoke alarms in 1890. They invented a device that produced an audible alarm when air temperature increased beyond a predetermined limit, according to their patent documents. They built the local heat detector, which we'd describe as a stand-alone alarm today, for buildings that didn't have alarm systems connected to the local fire department or other monitors such as storage facilities or buildings outside city limits. Their invention worked, but a string of fatal fires in the United States suggested that smoke detectors could warn of hazardous conditions more quickly than heat detectors. A smoldering fire, for example, may produce significant toxic smoke before creating enough heat to activate the Upton-Dibble Portable Electric Alarm.

Mechanical smoke detection started in the 1930s. Walter Jaeger, a Swiss physicist, was building a sensor for poisonous gas in 1930 when he stumbled into the next break-through in smoke detection.

His hypothesis that poisonous gas would ionize air in the sensor was wrong, but when he lit a cigarette to ponder his efforts, he saw the current in his meter change. Although poison gas didn't ionize the air, smoke did. The results of his experiment led other inventors to create smoke detectors.

In the 1950s, builders, residents and other consumers installed fire detectors that reacted to heating, but research into fatal fires such as the deadly 1958 inferno at Our Lady of the Angels Catholic School in Chicago suggested smoke detectors reacted more quickly to fire conditions than heat detectors. Early smoke detectors required high voltage and were expensive enough that they were inaccessible to most Americans outside of large industries.

Duane Pearsall and Stanley Peterson developed single-station photoelectric smoke alarms in 1965. They began manufacturing the first home smoke detectors in 1975 under the name Statitrol Corporation. Because they were powered by batteries, their smoke alarms could be maintained by anyone. The devices also were small—about the size of an adult's hand—and relatively inexpensive, which made them accessible to consumers.

The company grew quickly as it rushed to meet the demand created by changes to the nation's building, residential and fire codes that required smoke alarms in all one- and two-family dwellings. The Federal Housing Administration and Veterans Administration, as well as other federal agencies involved in mortgage lending, also required smoke alarms in homes to qualify for their funding. By 1980, over half of the nation's homes had at least one smoke alarm, suggesting education effectively spurred regulatory agencies (enforcement agencies) to adopt an engineering improvement. Four years later, 75 percent of homes had at least one smoke alarm. Engineering, Enforcement and Education, now joined by economic incentives and empowerment, were reducing risk.

Code officials met again and in 1984 began requiring smoke alarms in each level of a house, and four years later required alarms in new construction to be inter-connected so that if one activated, all would announce the alarm. The 1988 codes also required smoke alarms in all sleeping rooms.

The homes in which smoke alarms are not present account for 39 percent of reported home fires and half of reported home fire deaths. Of the homes that have smoke alarms, one fifth of those homes don't have any working smoke alarms, which may give those residents a false sense of security. Replacing batteries typically resolves the matter. Replacing alarms also is necessary because smoke alarms expire after ten years. Since 1995 smoke alarms have been manufactured with 10-year lithium batteries, which address both the life span of an alarm and the tedious need for annual battery maintenance.

There are two types of smoke alarms: photoelectric and ionization. Photoelectric alarms react more quickly to smoldering fires because the larger particles of soot reflect light within the detector's chamber and cause the alarm to activate. Ionization alarms react more quickly to flaming fires. Those alarms contain a tiny amount of radioactive material—often americium-241—that slowly sheds particles as it decays. Those particles remove electrons from air molecules within the chamber, leaving positive ions, while the freed electrons attach to other neutral molecules and form negative ions. Electrically-charged plates attract those ions creating a steady current. When smoke particles enter the chamber, they interrupt the current and activate the alarm.

Because both smoldering and flaming fires are possible in homes and other structures, having a combination of both alarms is the best way to detect smoke and protect occupants from that toxic, heated gas. Simply having smoke alarms isn't enough. Homeowners or other occupants of a structure must be responsible for replacing batteries and replacing the entire unit as needed.

Residential Sprinklers

Technology has existed for decades that not only detects fire, but sprays water to contain it. This technology lengthens the amount of time for a family's escape, protects valuable belongings from smoke, flames, and suppression efforts, improves the working conditions for responding firefighters, keeps the local economy intact, and reduces pollution. Yet residential sprinklers are demonized by some homebuilders, policy makers, and water purveyors, and misrepresented by Hollywood and other media.

Contrary to myth, sprinkler heads operate independently because each has a temperature-sensitive element. When the heat around a sprinkler head reaches a specific temperature such as 135 - 165 degrees Fahrenheit in most residential settings, either a glass bulb bursts or soldered pieces of metal separate, which allows water to flow from a pipe onto the fire. Most of the time, one or two sprinkler heads contain or extinguish the fire. Cigarette smoke, cooking vapors, and steam do not cause a sprinkler to activate accidentally. A residential sprinkler system is similar to having a fire engine built into the home—without having an additional four house guests or needing a larger garage—and can be the difference between life and death in rural, suburban, and urban communities.

Sprinklers are "green" technology, conserving water and reducing greenhouse gas emissions. A 10-year study in Scottsdale, Arizona, found that an average sprinkler system discharged 341 gallons of water on a fire while firefighter hoses averaged 2,935 gallons of water on similarly sized fires. Another recent study determined that a residential sprinkler system reduces the amount of carbon and toxic gases released into the atmosphere by as much as 98 percent.

Scottsdale isn't the only community that has documented the success of residential sprinklers in saving lives and preventing property damage. A lesser known but no less impressive report was written about fires in Bucks County, Pennsylvania, in 2011.

This report compares house fires with similar preconditions and different outcomes. For example, the writers picked a pair of house fires that occurred during the day where the occupants, each of whom had a medical condition that impaired her mobility, lived alone. A 53-year-old woman with multiple sclerosis in Doylestown called 911 in January 2008 to report her bed, where she was laying, was on fire. Police arrived two minutes after her call and firefighters a couple minutes later but none was able to rescue her. The fire, ignited by a cigarette, destroyed the house. "There were several instances where we went to the house [for a bed on fire] previously," James Donnelly, Doylestown's police chief, told a local newspaper reporter.

Compare that outcome to the experience of a 74-year-old woman in nearby Warwick Township. She was asleep in her bed when the bedding and padded headboard ignited. Heat from that fire activated a sprinkler head. Its water not only awakened the woman, but also extinguished the fire, and the system notified 911 dispatchers. She was scared and wet, according to the report's authors, but unscathed. Damage was limited to the room of origin.

The fatal fire caught the attention of local media for several days. The fire halted by the sprinkler system only netted a brief mention in the local newspaper. Life-saving technology not only exists but also works as designed. It receives little to no fanfare. In fact it is raked through the mud as unnecessary and luxurious equipment. As a life safety educator, I'm frustrated that sprinkler systems are opposed, but it comes down to risk perception.

Unfortunately, mandates requiring homebuilders to utilize this life-saving technology face strong opposition from builders, who cite added costs, threats to the local economy and aesthetics, from water purveyors who cite added costs and water scarcity, and even from fire officers and firefighters who seem nervous about losing their jobs if fewer fires occur. Across the nation, elected officials considering mandatory sprinklers face pitchfork-, lawsuit- and campaign-donation-wielding lobbyists as energetic as those who oppose restrictions on gun ownership and reproductive rights.

California and Maryland have adopted the residential sprinkler requirement for all new home construction. Colorado hasn't adopted that provision of the 2012 International Codes, but as Governor Hickenlooper said regarding wildfire preparedness, the state shouldn't be "dictating or usurping the local role." On the other end of the spectrum, Texas and many other states' legislators yield to the opposition and use preemption laws to prevent local communities from requiring residential sprinkler systems at all. Those states have taken away the personal responsibility and local control with their big-government-knows-best approach.

While residential sprinkler systems are just "some extra pipes with water," as South Metro Fire Rescue's Deputy Chief Mike Dell'Orfano often quips, they also are life-saving technology. A handful of agencies around the nation offer tax rebates to residents who retrofit their existing homes with sprinklers or add sprinklers to the designs of new homes to incentivize the use of this technology.

Residential, commercial and industrial sprinklers systems don't prevent fires; they react to them in order to reduce the impact of the fire to people and property. Hardening a home and landscape against wild-fire is a similar strategy in that its benefits are apparent only after a fire starts. Such hardening can occur without losing the natural appearance of homes. A company in California created a fire resistant wood-shake shingle roof by designing a fire-resistant roof assembly under the otherwise flammable shingles. Adding a sprinkler system, replacing a roof or any other technological improvement to a property is expensive, but those immediate costs can hide the benefits of preventing potential future costs. However, short-term costs can generate big returns.

These technologies contribute to the hózhǫ strategy, helping humans adapt to fire rather than ignore or fight it. Hopefully the benefits of saving lives—and actual economic benefits—will overcome marginal and mythic costs of new technologies and mitigation actions someday. Otherwise, America—its economy, its environment and we citizens—will keep burning. ⌒

Le-che-che,
the Hummingbird

Origins of Fire: Ancient Myths

*W*hen I was growing up I greatly enjoyed reading Rudyard Kipling's *Just So Stories*, recounting the colorful and fanciful ways various animals came to be the way they are. When I later went to college, Native American literature and history courses revived my interest in ancient stories that describe how and why our world is the way it is. Later, when I began researching new strategies for teaching our kids about fire, I turned to the fire myths related here.

Curiously, most of these fire myths have recurring themes, even though they originate within diverse cultures separated by vast geographical distances. One theme is fire's representation of divine power, over and above its practical uses; another is the strong presence within the myths of animal spirits possessing magical powers.

More compellingly, almost all of these fire myths share the same storyline, in which a distant tribe or a divine being possesses fire exclusively. Those who don't have fire—human or otherwise—conspire to take it through some form of trickery, suggesting they respect and even fear the guardians of fire enough not to confront them directly. Instead they rely on stealth and legerdemain. Once fire is obtained, the victorious culture then shares it with their neighbors in the ecosystem, after which fire is available to all humans.

This section contains the complete stories about the origins of fire mentioned in the chapters; their citations can be found in Research Notes.

How Fire Came to the Six Nations

– Kanien'kehake (Mohawks) –

Upstate New York and Surrounding States

Three Arrows was a boy of the Mohawk tribe. Although he had not yet seen fourteen winters he was already known among the Iroquois for his skill and daring. His arrows sped true to their mark. His name was given him when with three bone-tipped arrows he brought down three flying wild geese from the same flock.

He could travel in the forest as softly as the south wind and he was a skillful hunter, but he never killed a bird or animal unless his clan needed food. He was well-versed in woodcraft, fleet of foot, and a clever wrestler. His people said, "Soon he will be a chief like his father."

The sun shone strong in the heart of Three Arrows, because soon he would have to meet the test of strength and endurance through which the boys of his clan attained manhood. He had no fear of the outcome of the dream fast which was so soon to take.

His father was a great chief and a good man, and the boy's life had been patterned after that of his father. When the grass was knee-high, Three Arrows left his village with his father.

They climbed to a sacred place in the mountains. They found a narrow cave at the back of a little plateau. Here Three Arrows decided to live for his few days of prayer and vigil. He was not permitted to eat anything during the days and nights of his dream fast.

He had no weapons, and his only clothing was a breechclout and moccasins. His father left the boy with the promise that he would visit him each day that the ceremony lasted, at dawn.

Three Arrows prayed to the Great Spirit. He begged that soon his clan spirit would appear in a dream and tell him what his guardian animal or bird was to be. When he knew this, he would adopt that bird or animal as his special guardian for the rest of his life.

When the dream came he would be free to return to his people, his dream fast successfully achieved. For five suns Three Arrows spent his days and nights on the rocky plateau, only climbing down to the little spring for water after each sunset. His heart was filled with a dark cloud because that morning his father had sadly warned him that the next day, the sixth sun, he must return to his village even if no dream had come to him in the night.

This meant returning to his people in disgrace without the chance of taking another dream fast. That night Three Arrows, weak from hunger and weary from ceaseless watch, cried out to the Great Mystery.

"O Great Spirit, have pity on him who stands humbly before Thee. Let his clan spirit or a sign from beyond the thunderbird come to him before tomorrow's sunrise, if it be Thy will."

As he prayed, the wind suddenly veered from east to north. This cheered Three Arrows because the wind was now the wind of the great bear, and the bear was the totem of his clan. When he entered the cavern he smelled for the first time the unmistakable odor of a bear: this was strong medicine.

He crouched at the opening of the cave, too excited to lie down although his tired body craved rest. As he gazed out into the night he heard the rumble of thunder, saw the lightning flash, and felt the fierce breath of the wind from the north.

Suddenly a vision came to him, and a gigantic bear stood beside him in the cave. Then Three Arrows heard it say, "Listen well, Mohawk. Your clan spirit has heard your prayer. Tonight you will learn a great mystery which will bring help and gladness to all your people."

A terrible clash of thunder brought the dazed boy to his feet as the bear disappeared. He looked from the cave just as a streak of lightning flashed across the sky in the form of a blazing arrow. Was this the sign from the thunderbird? Suddenly the air was filled with a fearful sound. A shrill shrieking came from the ledge just above the cave. It sounded as though mountain lions fought in the storm; yet Three Arrows felt no fear as he climbed toward the ledge.

As his keen eyes grew accustomed to the dim light he saw that the force of the wind was causing two young balsam trees to rub violently against each other. The strange noise was caused by friction, and as he listened and watched, fear filled his heart, for, from where the two trees rubbed together a flash of lightning showed smoke. Fascinated, he watched until flickers of flames followed the smoke.

He had never seen fire of any kind at close range nor had any of his people. He scrambled down to the cave and covered his eyes in dread of this strange magic. Then he smelt bear again and he thought of his vision, his clan spirit, the bear, and its message. This was the mystery which he was to reveal to his people. The blazing arrow in the sky was to be his totem, and his new name—Blazing Arrow.

At daybreak, Blazing Arrow climbed onto the ledge and broke two dried sticks from what remained of one of the balsams. He rubbed them violently together, but nothing happened. "The magic is too powerful for me," he thought. Then a picture of his clan and village formed in his mind, and he patiently rubbed the hot sticks together again.

His will power took the place of his tired muscles. Soon a little wisp of smoke greeted his renewed efforts, then came a bright spark on one of the sticks. Blazing Arrow waved it as he had seen the fiery arrow wave in the night sky. A resinous blister on the stick glowed, then flamed—fire had come to the Six Nations!

Who Was Given the Fire

– Cowichan –

Koksilah and Cowichan River Watersheds,
West Coast of British Columbia

*O*ur fathers tell us that very long ago our people did not know the use of fire. They had no need for fire to warm themselves, because they lived in a warm country. They ate their meats raw or dried by the sun. But after a while their climate grew colder. They had to build houses for shelter, and they wished for something to warm their homes.

One time when a number of them were seated eating an animal they had just found in one of their pits, a pretty bird came and fluttered above their heads. It seemed to be either watching them or looking for a share in the meat. Seeing the bird flying about, some people tried to kill it. Others, more kind, said, "Little bird, what do you want?"

"I know your needs," the bird replied, "and I have come to you, bringing the blessings of fire."

"What is fire?" asked all of them.

"Do you see that little flame on my tail?" asked the bird.

"Yes," all answered.

"Well, that is fire. Today each of you must gather a small bunch of pitch wood. With it you can get fire from the flame on my tail. Tomorrow morning I will come here early. Every one of you will meet me here, bringing your pitch wood with you."

Early next morning all arrived at the chosen place, where the bird was awaiting their coming.

"Have you brought your pitch wood?" asked the bird.

"Yes," replied all of the people.

"Well, then," said the bird, "I am ready. But before I go, let me tell you the rules. None of you can obtain my fire unless you obey the rules. You must be persevering, and you must do good deeds. You must strive for the fire, in order that you may think more of it. And none need to expect to get it who has not done some good deed.

"Whoever comes up with me," continued the bird, "and puts his pitch wood on my tail, he will have the fire. Are you all ready?"

"Yes," replied everyone.

Away flew the bird, followed by all the people, young and old, men and women and children. Helter-skelter they ran, over rocks and fallen timber, through swamp and stream, over prairies and through forests. Some of them got hurt. Others peeled their shins as they fell off the rocks and stumbled over the logs. Many people splashed through mud and water. Others were badly scratched and had their clothes torn among the bushes. Many turned and went home, saying, "Anything so full of danger is not worth trying for." Other people became so weary they gave up. But the bird kept on.

At last a man came up to it, saying, "Pretty bird, give me your fire. I have kept up with you, and I have never done anything bad."

"That may be true," replied the bird. "But you cannot have my fire because you are too selfish. You care for nobody as long as you yourself are right."

So away flew the bird.

After a while another man came up, saying, "Pretty bird, give me your fire. I have always been good and kind."

"Perhaps you have been," answered the bird. "But you cannot have my fire because you stole your neighbors' wife."

So the bird flew away again. By this time few people followed it, most of them having given up the chase.

At last the bird came to where a woman was taking care of a poor, sick old man. It flew straight to her and said, "Bring your pitch wood here and get the fire."

"Oh, no," said the woman. "I cannot do so because I have done nothing to deserve it. What I am doing is only my duty."

"Take the fire," said the bird. "You are welcome to it. It is yours, for you are always doing good and thinking it only your duty. Take the fire and share it with the other people."

So the woman put her pitch wood on the bird's tail and got the fire. Then she gave some to all the others, and people have never since been without it. Fire has cooked their food and warmed their lodges. That is how, in the long, long ago, the Cowichan first got fire.

Bobok and the First Fire

– Yaqui –

*Yaqui River Watershed, Southwestern United States &
Northwestern Mexico*

*N*ow there is fire in all rocks, in all sticks. But long ago there wasn't any fire in the world, and all of the Yaquis and the animals and the creatures of the sea, everything that lived, gathered in a great council in order to understand why there was no fire.

They knew that somewhere there must be fire, perhaps in the sea, maybe on some islands, or on the other side of the sea. For this reason, Bobok, the Toad, offered to go get this fire. The Crow offered to help him and also the Roadrunner and the Dog. These four, the winged animals and the dog, went along to help. But Bobok, the Toad, alone, knew how to enter the water of the sea and not die.

The God of Fire would not permit anyone to take his fire away. For this reason he still sends thunderbolts and lightning at anyone who carries light or fire. He is always killing them.

But Bobok entered the house of the God of Fire and stole the fire. He carried it in his mouth, traveling through the waters. Lightning and thunder made a great noise and many flashes. But Bobok came on, safe beneath the waters. Then there formed on the flooding water little whirlpools of water full of rubbish and driftwood.

Suddenly not only one toad was to be seen, but many swam in the waters, many, many toads. They were all singing and carrying little bits of fire. Bobok had met his sons and had given some fire to one, then another, until every toad had some. These carried fire to the land where they were awaited by the Dog, the Roadrunner, and the Crow. Bobok gave his fire to those who could not enter the water.

The God of Fire saw this and threw lightning at the Crow and the Roadrunner and the Dog. But many toads kept on coming and bearing fire to the world. These animals gave light to all the things in the world. They put it into sticks and rocks. Now men can make fire with a drill because the sticks have fire in them.

～∙⁌

Koo-loo'-loo Fetches Fire from the Star-Women

– Me-wuk (Miwok) –

Northern California

The first fire was made by the Doctor Birds at the birth of Wek'-wek. The next fire was made by Ke'-lok the North Giant. After Ke'-lok's death and after his fire had burnt up the world and had burnt itself out, there was no fire except that of the Hul-luk mi-yum'-ko, the Star-women, which was close by the elderberry tree, way off in the east where the Sun gets up.

O-let'-te said to his grandson, Wek'-wek: "Now we have people, and elderberry music for the people, but we have no fire for them to cook with; the Star-women have it; we must steal it."

"How?" asked Wek'-wek.

"Send Koo-loo'-loo the Humming-bird; he is faster than you. Tell him to catch a little spark and bring it quickly," replied O-let'-te.

"All right," answered Wek'-wek, and he sent Koo-loo'-loo to fetch the fire. Koo-loo'-loo shot out swiftly and soon reached the Star-women by the elderberry tree in the far east, in the place where the Sun gets up. Here he hid and watched and waited, and when he saw a little spark of fire, he darted in and seized it and brought it back quickly to Wek'-wek and O-let'-te. He held it tight under his chin, and to this day if you look under the Humming-bird's chin you will see the mark of the fire.

Then Wek'-wek asked: "Where shall we put it?"

O-let'-te answered, "Let us put it in *oo'-noo*, the buckeye tree, where all the people can get it." So they put it in *oo'-noo*, the buckeye tree, and even now whenever an Indian wants fire he goes to the *oo'-noo* tree and gets it.

~~~

Tol-le-loo and the Flute

– Me-wuk (Miwok) –

Northern California

*L*ong before the Alisal rancheria was established, the Valley People lived in California's San Joaquin Valley, about a day's walk from the eventual site of Alisal and not far from the present town of Stockton. Their chiefs were Wek-wek, the Falcon, and We-pi-ah-gah, the Golden Eagle.

Their neighbors to the east, the Mountain People, lived in darkness in the Sierra Nevada mountains. Although, they wanted fire, the Mountain People did not know where or how to obtain it. O-la-choo, the Coyote-man, tried to find it but failed. Eventually, Tol-le-loo, the White-footed Mouse, discovered that the Valley People had fire, and O-la-choo sent him to steal it.

Taking his elderberry flute with him, Tol-le-loo traveled west until he reached the homes of the Valley People. Arriving outside their roundhouse, Tol-le-loo sat down and began to play his flute. Finding the music pleasant to listen to, the Valley People invited Tol-le-loo to come inside and continue his playing. Soon all the people began to feel sleepy. Now Wit-tah-bah the Robin was pretty sure that Tol-le-loo was planning on stealing their fire, so he spread himself over the embers to protect it. And that is why the robin's is breast is red today.

Tol-le-loo kept playing his flute, and pretty soon everyone, including Wit-tah-bah the Robin, had fallen asleep. Seizing this opportunity, Tol-le-loo ran up to the sleeping Wit-tah-bah, and cut a small hole in his wing. Then he crawled through the hole and placed the fire inside his flute. Running out of the roundhouse, he climbed to the top of Mount Diablo, where he built a great fire that lit up the entire countryside, including the blue Sierra Nevada mountains to the east where the Mountain People lived.

When Wek-wek the Falcon awoke and saw the fire on Mount Diablo, he knew that Tol-le-loo had stolen the Valley People's fire. So he set out after Tol-le-loo, and eventually caught him. Tol-le-loo denied having taken the fire, and told Wek-wek to search him if he doubted him. Wek-wek searched but could not find the fire because it was inside Tol-le-loo's flute. So Wek-wek tossed Tol-le-loo into some water and let him go on his way.

Tol-le-loo climbed out of the water, and continued east to the mountains, all the while carrying the fire in his flute. Arriving home, he took the fire out of the flute, and placed it on the ground. Then covering it with leaves and pine needles, he wrapped it up in a small bundle.

Le-che-che the Hummingbird and another bird went after it, but they could not catch it and returned empty-handed.

O-la-choo the Coyote-man could smell the fire, and wanted to steal it. He approached the bundle, and pushed it with his nose, preparing to swallow it. Suddenly, however, the fire shot up into the sky and became the Sun.

The people took the fire that was left and put it into two trees, the buckeye and the incense cedar, where legend says it still resides. From that time on, the Mountain People made their fire drills from the wood of these two trees.

Maui and the Mud-hen

– Polynesia –

Hawai'an Islands

*H*ina, Maui's mother, wanted fish. One morning early Maui saw that the great storm waves of the sea had died down and the fishing grounds could be easily reached. He awakened his brothers and with them hastened to the beach. This was at Kaupo on the island of Maui. Out into the gray shadows of the dawn they paddled.

When they were far from shore they began to fish. But Maui, looking landward, saw a fire on the mountain side.

"Behold," he cried. "There is a fire burning. Whose can this fire be?"

"Whose, indeed?" his brothers replied.

"Let us hasten to the shore and cook our food," said one.

They decided that they had better catch some fish to cook before they returned. Thus, in the morning, before the hot sun drove the fish deep down to the dark recesses of the sea, they fished until a bountiful supply lay in the bottom of the canoe.

When they came to land, Maui leaped out and ran up the mountainside to get the fire. For a long, long time they had been without fire. The great volcano Haleakala above them had become extinct and they had lost the coals they had tried to keep alive. They had eaten fruits and uncooked roots and the shell fish broken from the reef—and sometimes the great raw fish from the far-out ocean. But now they hoped to gain living fire and cooked food.

But when Maui rushed up toward the cloudy pillar of smoke he saw a family of birds scratching the fire out. Their work was finished and they flew away just as he reached the place.

Maui and his brothers watched for fire day after day—but the birds, the curly-tailed Alae (mudhens) made no fire. Finally the brothers went fishing once more—but when they looked toward the mountain, again they saw flames and smoke. Thus it happened to them again and again.

Maui proposed to his brothers that they go fishing, leaving him to watch the birds. But the Alae counted the fishermen and refused to build a fire for the hidden one who was watching them. They said among themselves, "Three are in the boat and we know not where the other one is, we will make no fire today."

So the experiment failed again and again. If one or two remained or if all waited on the land there would be no fire—but the dawn which saw the four brothers in the boat, saw also the fire on the land.

Finally Maui rolled some kapa cloth together and stuck it up in one end of the canoe so that it would look like a man. He then concealed himself near the haunt of the mud-hens, while his brothers went out fishing. The birds counted the figures in the boat and then started to build a heap of wood for the fire.

Maui was impatient—and just as the old Alae began to select sticks with which to make the flames he leaped swiftly out and caught her and held her prisoner. He forgot for a moment that he wanted the secret of fire making. In his anger against the wise bird his first impulse was to taunt her and then kill her for hiding the secret of fire.

But the Alae cried out: "If you are the death of me, my secret will perish also, and you cannot have fire."

Maui then promised to spare her life if she would tell him what to do.

Then came the contest of wits. The bird told the demi-god to rub the stalks of water plants together. He guarded the bird and tried the plants. Water instead of fire ran out of the twisted stems. Then she told him to rub reeds together—but they bent and broke and could make no fire. He twisted her neck until she was half dead—then she cried out: "I have hidden the fire in a green stick."

Maui worked hard, but not a spark of fire appeared. Again he caught his prisoner by the head and wrung her neck, and she named a kind of dry wood. Maui rubbed the sticks together, but they only became warm. The neck-twisting process was resumed and repeated again and again, until Maui had tried every tree and the mud hen was almost dead. At last Maui found fire. Then as the flames rose he said: "There is one more thing to rub." He took a fire stick and rubbed the top of the head of his prisoner until the feathers fell off and the raw flesh appeared. Thus the Hawaiian mud-hen and her descendants have ever since had bald heads, and the Hawaiians have had the secret of fire making.

Mantis Fools Greedy Ostrich

– San –

Kalahari Desert, Botswana

*O*ne day, Mantis smelled a wonderful aroma floating through the countryside. Curious, Mantis peeked through a bush and saw Ostrich roasting food over a fire.

When Ostrich finished eating, he took the fire and tucked it under his wing. Mantis had never seen fire, and he now wanted it for himself.

When Ostrich jogged by, Mantis called out to him and told him of a wonderful tree filled with fruit. Excited, Ostrich followed Mantis to a tree covered with yellow plums.

"The best fruit," said Mantis, "is at the top."

Ostrich eagerly reached up with his long neck and extended his wings to keep his balance. As soon as Ostrich opened his wings, Mantis snatched the fire and fled. Since then, Ostrich has kept its wings at its side and has never attempted to fly.

The Theft of Fire from Thunder

– Maidu –

Northern California

*A*t one time the people had found fire, and were going to use it; but Thunder wanted to take it away from them, as he desired to be the only one who should have fire. He thought that if he could do this, he would be able to kill all the people.

After a time he succeeded and carried the fire home with him, far to the south. He put Woswosim (a small bird) to guard the fire, and see that no one should steal it. Thunder thought that people would die after he had stolen their fire, for they would not be able to cook their food; but the people managed to get along...

Mouse, Deer, Dog, and Coyote were the ones who were to try, but all the other people went too. They took a flute with them for they meant to put the fire in it.

They traveled a long time, and finally reached the place where the fire was. They were within a little distance of Thunder's house, when they all stopped to see what they would do...

After a while Mouse was sent up to try and see if he could get in. He crept up slowly till he got close to Woswosim, and then saw that his eyes were shut. When Mouse saw that the watcher was asleep, he crawled to the opening and went in. Thunder had several daughters, and they were lying there asleep.

Mouse stole up quietly, and untied the waist-string of each one's apron, so that should the alarm be given, and they jump up, these aprons or skirts would fall off, and they would have to stop to fix them. This done, Mouse took the flute, filled it with fire, then crept out, and rejoined the other people who were waiting outside.

For a while all went well, but when they were about half-way back, Thunder woke up, suspected that something was wrong, and asked, "What is the matter with my fire?"

He jumped up with a roar of thunder, and his daughters were thus awakened, and also jumped up; but their aprons fell off as they did so, and they had to sit down again to put them on. After they were all ready, they went out with Thunder to give chase. They carried with them a heavy wind and a great rain and a hailstorm, so that they might put out any fire the people had. Thunder and his daughters hurried along, and soon caught up with the fugitives, and were about to catch them, when Skunk shot at Thunder and killed him.

Then Skunk called out, "After this you must never try to follow and kill people. You must stay up in the sky, and be the thunder. That is what you will be." The daughters of Thunder did not follow any farther; so the people went on safely, and got home with their fire, and people have had it ever since.

The Flood and the Theft of Fire

– Tolowa –

Southwestern Oregon

A long time ago there came a great rain. It lasted a long time and the water kept rising till all the valleys were submerged, and the [ancestral Tolowa] tribes fled to the high lands.

But the water rose, and though the Indians fled to the highest point, all were swept away and drowned—all but one man and one woman. They reached the very highest peak and were saved. These two Indians ate the fish from the waters around them.

Then the waters subsided. All the game was gone, and all the animals. But the children of these two Indians, when they died, became the spirits of deer and bear and insects, and so the animals and insects came back to the Earth again.

The Indians had no fire. The flood had put out all the fires in the world. They looked at the moon and wished they could secure fire from it. Then the Spider Indians and the Snake Indians formed a plan to steal fire. The Spiders wove a very light balloon, and fastened it by a long rope to the Earth. Then they climbed into the balloon and started for the moon. But the Indians of the Moon were suspicious of the Earth Indians. The Spiders said, "We came to gamble." The Moon Indians were much pleased and all the Spider Indians began to gamble with them. They sat by the fire.

Then the Snake Indians sent a man to climb up the long rope from the Earth to the moon. He climbed the rope, and darted through the fire before the Moon Indians understood what he had done. Then he slid down the rope to Earth again. As soon as he touched the Earth he traveled over the rocks, the trees, and the dry sticks lying upon the ground, giving fire to each. Everything he touched contained fire. So the world became bright again, as it was before the flood.

When the Spider Indians came down to Earth again, they were immediately put to death, for the tribes were afraid the Moon Indians might want revenge.

~

The Boy Who Brought Fire

– Nimipu (Nez Perce) –

Columbia River Plateau in Washington, Oregon & Idaho

*L*ong ago, the Nimipu had no fire. They could see fire in the sky sometimes, but it belonged to the Great Power. He kept it in great black bags in the sky. When the bags bumped into each other, there was a crashing, tearing sound, and through the hole that was made fire sparkled.

People longed to get it. They ate fish and meat raw as the animals do. They ate roots and berries raw as the bears do. The women grieved when they saw their little ones shivering and blue with cold. The medicine men beat on their drums in their efforts to bring fire down from the sky, but no fire came.

At last a boy just beyond the age for the sacred vigil said that he would get the fire. People laughed at him. The medicine men angrily complained, "Do you think that you can do what we are not able to do?"

But the boy went on and made his plans. The first time that he saw the black fire bags drifting in the sky, he got ready. First he bathed, brushing himself with fir branches until he was entirely clean and was fragrant with the smell of fir. He looked very handsome.

With the inside bark of cedar he wrapped an arrowhead and placed it beside his best and largest bow. On the ground he placed a beautiful white shell that he often wore around his neck. Then he asked his guardian spirit to help him reach the cloud with his arrow.

All the people stood watching. The medicine men said among themselves, "Let us have him killed, lest he make the Great Power angry."

But the people said, "Let him alone. Perhaps he can bring the fire down. If he does not, then we can kill him."

The boy waited until he saw that the largest fire bag was over his head, growling and rumbling. Then he raised his bow and shot the arrow straight upward. Suddenly, all the people heard a tremendous crash, and they saw a flash of fire in the sky. Then the burning arrow,

168

like a falling star, came hurtling down among them. It struck the boy's white shell and there made a small flame.

Shouting with joy, the people rushed forward. They lighted sticks and dry bark and hurried to their tipis to start fires with them. Children and old people ran around, laughing and singing.

When the excitement had died down, people asked about the boy. But he was nowhere to be seen. On the ground lay his shell, burned so that it showed the fire colors. Near it lay the boy's bow. People tried to shoot with it, but not even the strongest man and the best with bow and arrow could bend it.

The boy was never seen again. But his abalone shell is still beautiful, still touched with the colors of flame. And the fire he brought from the black bag is still in the center of each tipi, the blessing of every home.

Stealing Fire from the Bandicoot

– Aborigine –

Victoria, Australia

The bandicoot was once the sole owner of fire, and cherishing his fire-brand, which he carried with him wherever he went. He obstinately refused to share the flame with anyone else. Accordingly the other animals held a council and determined to get fire either by force or by stratagem, deputing the hawk and the pigeon to carry out their purpose. The latter, waiting for a favorable moment when he thought to find it unguarded, made a dash for it; but the bandicoot saw him in time, and seizing the brand, he hurled it toward the river to quench it. The sharp eyes of the hawk saw it falling, and swooping down, with his wing he knocked it into the long dry grass, which was thus set alight so that the flames spread far and wide, and all people were able to procure fire.

Biliku Saves the Fire

– Aka-Jˇeru –

Andaman Islands, Bay of Bengal

In the days of the ancestors they had no fire. Biliku had fire. While Biliku slept Maia Lirčitmo (Sir Kingfisher) came and stole fire. As he was taking the fire Biliku awoke and saw him. Lirčitmo swallowed the fire. Biliku took a pearl shell and threw it at Lirčitmo and cut off his head. The fire came out (of his neck). The ancestors got the fire. Lirčitmo became a bird.

The First Bush Fire

– Aborigine –

Australia

There was a time when the Australian bush was different from what it is today. Trees were bigger and their wood softer. There were more and bigger and brighter flowers. And the land—especially the mountains—was far more densely clothed with verdure. But some change came, and it was not good for the land. Seeds failed to germinate, and where fertile tracts had been now desert appeared.

Somewhere away in the south, perhaps away over in Victoria, there lived a great king. His people were very numerous, for he had imposed his will upon other tribes than that which was his when he was first made ruler, and he had succeeded in welding them all together into one harmonious group. They revered him and all sorts of presents were laid by them at his feet.

Yet he never shirked work, and he took a place amongst the hunters just the same as if he were not a king. He must have come as far north as the Burragorang—if, indeed, he had not come further—for the Hunter River people have a story just like this.

Living in a valley between two mountains was a very small tribe—an unusually docile people. They were an offshoot of that tribe who owned the country at the head of Cox's River. The Powerful chief heard of them, and he determined to find out what they were like and add them to his subjects. So he set out by stealth.

Wrapping himself about with a wombat skin he came to the hiding-place. He was a very big man and he could not well conceal himself in so small a skin as that of a wombat—even the biggest of them. Therefore when he was within sight of the camp he hid behind a rock. He saw that the tribe was very busy just at the time cooking game by heating stones and placing them one after another around and upon the carcasses. The handling of the stones was made easy by the wrapping of *waratah* stems about the fingers.

This wrapping of *waratah* stems to make a person immune from burns was so believed in by the blacks that they came to the first blacksmiths that they saw and offered them the twigs, indicating that if they would wear them no flying sparks could injure them.

Now amongst the people was one maiden of exceptional beauty. Some say that the reason for these blacks ostracizing themselves was that many years before a beautiful woman wished her pretty baby to be called Krubi, and the other Krubi was not yet old. So the mother gathered her children about her and went away, and the family increased and increased, and always there were those quite beautiful enough to be called by the coveted name of Krubi.

But again it might have been because of the social system as explained in another legend that any portion of a tribe went further afield and formed another and distinct group.

Anyhow, when the king saw this maiden he lost all his cunning, so entrancing was she, and jumping up without reserve he ran towards the people. They started up in fear and scattered in many directions. The king called to them not to fear him, but she did not understand his speech. The beautiful maiden soon found that the stranger pursued her, and her alone. She was very fleet of foot and very cunning, and by dodging and crouching she eluded him. Sometimes she was so close and so still, standing beside a tree, that he ran past her, and only by not hearing her crashing through the bushes and stamping on the twigs and leaves did he know that he had gone too far. No sooner did he turn than she bounded off again.

There was a stream clattering down a gully and falling over boulders and ledges into pure cold pools, and towards that stream the girl now ran. She knew of some footholds close to a waterfall, and, indeed, sometimes even behind it, that led to a very large and deep pool, and outstripping the now panting king she reached it. Never hesitating she clambered and swung herself down, and reaching the bottom she swung over the last ledge and slipped into the water.

The king reached the top of the fall, and believing that his quarry could not have gone down there he retraced his steps. He went right back to the camping place and found it, of course, silent and deserted. So he returned to the people he had left and told nothing of where he had been and what he had seen.

As often as he could he went back, and though sometimes he saw some of the people, never did he catch sight of the girl. He went so often that the others grew not to fear him. They guessed his desire, and they aided the girl at all times to hide from him.

One day he overlooked the people who were unaware of the fact that from a neighboring prominence the king could see them. They had grown so careless that they did not think of pitching their *mia-mias* where no one could see them from a distance.

And he caught a sight of the maiden of his desires. But she saw him and once more she had to run as if for her life. He did not hurry after her. Instead he got two dry sticks, and sharpening one on a stone he placed the flatter of the two on the earth before him, and putting the point of the other on it he rubbed and twisted it round and back between his palms until he had caused a fire to glow. He had dry ferns and grass ready, and placing them on the glowing spot he gently blew until the flame burst out. He added more fuel until he had a big blaze. The wind blew in the direction of the little tribe, and soon a great roaring fire was leaping and leaping and shooting out curling masses of consuming flame.

The girl saw it coming. The tribe saw it also. Away they all ran, bounding and crashing, but the fire came faster. It overtook some of them and they perished. On the blackened cleared ground the now wicked king followed. But he could not go fast. The smoldering sticks and rubbish were still hot. There were no *waratahs* growing just here for the *waratah* does not grow as profusely as, say, the gums, but in patches far apart. Hot as he was and suffering burns as he was still, he examined everybody he came across, but they were none of them the one he sought. A last he came to two little heaps of clay. What was this? These heaps told to him a story. They were fresh. They were composed of the clay that the tribal doctor used to make the mystic markings, and the tribal priest used for the same purpose when he wished to invoke the aid of a Great Spirit. Who had used it? Not once had he seen anyone who looked like one who had been initiated into the doubly-hidden mysteries of the rite that gave power to invoke the Spirit. Surely the girl had not seen that corroboree. If she had, then not only could he never capture her, but he was himself lost.

And lost he surely was. For on looking behind him he found that almost as the fallen seeds of the trees were being consumed by fast or slow smoldering, they were bursting with new life, and plants were springing up in such profusion as to block his view.

In what direction had he come? Which way would he turn to go back? The smoke was so dense that he could not see the sun. The trees that lay over from the prevailing winds and gave some idea of direction were burning, and their small branches were gone.

Surely, he thought, this is the work of the maiden and she knew more than any woman was allowed to know.

He wandered on and on, the bush growing denser. He stayed sometimes to pick up something to eat, for burned and roasted game lay in his path, and succulent roots were cooked. He wandered for many days quite lost.

The girl had visited at night the tribe from which her family was an offshoot, and had come across the corroboree that taught her how to paint herself, and this she had done, and the charm was hers.

A new camping place was chosen by the few who escaped that terrible fire, and the year rolled away. The young plants flowered and their seeds fell, but the next year no new plants came up. This was noticed and talked about by all the people. Even on the river where a few of her people were now living no seeds sent out the little *plumule* nor their little radicle, and no new plants grew to grace the world with fresh flowers, nor to produce the roots nor fruits for food.

Again the maiden thought of beseeching the spirit. She went back to the old ground all alone and she found the clay. She painted herself and awaited results. She heard the spirit and she talked with it. Then she noticed that just before her a little smoke wreath curled up into the air. Then a flame burst, and in a very little while a fierce bush fire was raging.

The girl was satisfied that a fire was what was needed and she sent word to the river to say that all would soon be well with the world. That the seeds would germinate and new plants would grow up and flower and all would be gay as before.

Since that time bush fires do not need any mystic markings nor special communings by special people. Limbs of trees rub themselves hot on dry days and make flame. The hot sun shining on the mica in the rocks set fire to the tiny mosses that are dried there. And so without human agency the fires come that are necessary to make our Australian seeds burst into the life of a new and growing plant.

The black people knew this, and they were well aware that the seeds must be burnt and so this knowledge gave rise to the legend written here.

How Coyote Brought Fire to the People

– Karuk (Karok) –

Klamath River Watershed, Northwestern California

In the beginning, the animal people had no fire. The only fire anywhere was on the top of a high, snow-covered mountain, where it was guarded by the skookums. The skookums were afraid that if the animal people had any fire they might become very powerful, as powerful as the skookums. So the skookums would not give any of the fire away to anyone.

Because the animal people had no fire, they were always shivering, and they had to eat their food raw. When Coyote came along he found them cold and miserable.

"Coyote," they begged, "you must bring us fire from the mountain or we will one day die of all this cold."

"I will see what I can do for you," promised Coyote. As soon as the sun came up the next morning, Coyote began the long and difficult climb to the top of the mountain where the skookums kept the fire.

When he got to the top he saw that three wrinkled, old skookums, all sisters, guarded the fire all day and all night, each taking a turn. While one kept watch the other two ate and slept in a lodge nearby. When it was time to change the watch the one at the fire would go to the door of the lodge and call out "Sister, get up and guard the fire."

At dawn, the skookum who had been watching the fire all night was always stiff with the cold and she walked very slowly through the snow to the lodge door to call her sister. "This is the time to steal a brand of the fire," thought Coyote to himself. But he knew, too, that he would be chased. And he knew that even though the skookums were old they were swift and strong runners. Coyote would have to devise a plan.

Coyote thought and thought, but he could not come up with any plan. So he decided to ask his three sisters who always lived in his stomach in the form of huckleberries to help him. They were very wise, and they would tell him what to do.

At first, Coyote's sisters were reluctant to help him. "If we tell you," they said, "you will only say that you knew it all along."

Coyote remembered that his sisters were afraid of hail and so he called up into the sky, "Hail! Hail! Fall down from the sky."

This made his sisters very afraid. "Stop!" they called. "Don't bring the hail down. We will tell you what you want to know."

Coyote's sisters then told him how to steal the fire and get it down the mountain to the people without getting caught. When they had finished talking, Coyote said, "Yes, that was my plan all along."

Coyote then went to see the animal people. He called everyone together, as his sisters had directed, and told each animal Antelope, Fox, Weasel, Beaver, Squirrel and the others to take up certain places along the mountainside. When they were all in place, they stretched in a long line from the top of the mountain all the way back to the village.

Coyote climbed back up the mountain and waited for sunrise. The old skookum who was watching the fire had keen eyes and she saw him. But she thought it was just an animal skulking around looking for scraps.

At dawn the skookum left the fire and walked slowly over to the lodge door. "Sister, get up and guard the fire."

Just at that moment Coyote sprang from the bushes. He seized a burning brand from the fire and ran away as fast as he could across the snow. The three skookums were right behind him in an instant. They were so close they were showering Coyote with the snow and ice they were churning up in their fury. Coyote was running as fast as he had ever run in his life. He leaped over cracks in the ice and rolled part way down the mountain like a snowball, but the skookums were right behind him. So close behind that their hot breath scorched his fur.

When Coyote finally reached the tree line, Cougar jumped out from his hiding place, snatched up the fire brand and raced away just as Coyote fell flat on his face from exhaustion. Cougar ran all the way to the high trees where he gave the fire to Fox. Fox raced until he came to the heavy undergrowth where he gave the fire to Squirrel. Squirrel ran away through the trees, leaping from branch to branch. The skookums could not go through the trees so they planned to catch Squirrel at the edge of the woods. But Antelope was waiting there to get the fire from Squirrel, and Antelope, who was the fastest of all the animals, bounded away across the meadow. One after another, each one of the animals carried the fire, but the skookums stayed right behind them.

Finally, when there was only a glowing coal left, the fire was passed to Frog. Frog swallowed the hot coal and hopped away as fast as he could hop. The skookums were almost on top of him when he dove into a deep river and swam across to the other side. The youngest skookum had already leaped across the water and was waiting for him. As soon as he landed, Frog saw what had happened and jumped between the skookum's legs and bounded away. An instant later the skookums were on him again and Frog was too tired to jump. So he spat the hot coal out on Wood and Wood swallowed it. The three skookums stood there not knowing what to do. None of them could figure a way to take the fire away from Wood. After a while they left and went slowly back to their lodge on the top of the mountain.

Coyote then called the animals together and they all gathered around Wood. Coyote, who was very wise, knew how to get the fire out of Wood.

He showed the animals how to rub two dry sticks together until sparks came. Then he showed them how to collect dry moss and make chips of wood to add to the sparks to make a little fire. Then he showed them how to add small twigs and pine needles to make a bigger fire.

From then on, the people knew how to get the fire out of Wood. They cooked their meat, their houses were warm, and they were never cold again.

Kanbi and Jitabidi Start the First Fire

– Aborigine –

Northern Coast of New South Wales, Australia

*L*ong ago, before even the Dreamtime, there was a tribe of people who did not live on the earth. They lived in the sky world and their camp was near the two brightest stars so that they could light their fire-sticks from them. They were the only people anywhere who had the use of fire. The people on earth had to manage without it.

As even in the sky world, life is not always perfect, there came a time when there was not enough food. Some of the most adventurous of the sky world people decided to come down to the earth world to hunt.

"We will hunt possums and collect nuts and berries. It won't take long and we can bring back enough food for everyone," they told their companions.

Two brothers, named Kanbi and Jitabidi, brought their fire-sticks down to earth with them and left them smoldering while they went off hunting. Hunting possums turned out to be a lot more difficult than they had expected and the time drew out and the land was very quiet. The fire-sticks became bored and began to play 'chase'. They were very clever at this game, running from place to place, and everywhere they touched the dry grass it caught alight. Gradually all the little fires grew together into one big fire and the smoke could be seen from a long way off. As soon as the sky brothers saw the smoke they left the hunting party and hurried back to put out the fires.

The Aboriginal people who lived in the area had also seen the smoke and had come to see what was going on. They had never seen fire before and at first they were very afraid. It did not take them long, however, to realize that this strange phenomenon could be extremely useful to them, providing them with light and warmth. They also noticed that some possums the sky brothers had caught had been cooked by the fire and smelled wonderful and savory. They realized that they too could make their food more tender and tasty with fire.

Before Kanbi and Jitabidi could finish putting out the fires, several of the Aboriginal men lit fire-sticks for themselves and hurriedly carried them back to their camps.

"We must watch over these fire-sticks and carefully keep them burning forever," they told one another.

Kanbi and Jitabidi quickly gathered up their playful fire-sticks and returned to their campsite in the sky. They were terribly afraid the earth people would inflict some punishment on them for having caused such a disturbance. But the earth people were in awe of their sky visitors and rather than being angry about the burnt grass were excited and grateful for the wonderful gift of fire.

Stealing Fire from the Women

– Aborigine –

New South Wales, Australia

According to this, fire was originally owned by two women (Kangaroo-Rat and Bronze-Winged Pigeon) who kept it concealed in a nutshell. For a long time the other animals could not discover how these women were able to cook their food; but at last they set spies to watch them and so learned the secret, whereupon, resolving to secure fire by a ruse, they arranged a dance and invited the two women to be present. One after another the different animals danced in ludicrous positions in an attempt to make the women laugh; and at length one performer succeeded so that the women, convulsed with merriment, rolled upon the ground. This was just what the conspirators had been waiting for, and rushing up, they seized the bag in which was the nut that contained the fire. Opening this and scattering the flame about, they set the grass alight, and in this way fire was caught in the trees, whence ever since it can be procured from their wood by means of friction.

Coyote Steals Fire

– Zia Pueblo –

New Mexico

𝒶long, long time ago, the people became tired of feeding on grass, like deer and wild animals, and they talked together how fire might be found. The Ti-amoni said, "Coyote is the best man to steal fire from the world below," so he sent for Coyote.

When Coyote came, the Ti-amoni said, "The people wish for fire. We are tired of feeding on grass. You must go to the world below and bring the fire."

Coyote said, "It is well, father. I will go."

So Coyote slipped stealthily to the house of Sussistinnako.

It was the middle of the night. Snake, who guarded the first door, was asleep, and he slipped quickly and quietly by. Cougar, who guarded the second door, was asleep, and Coyote slipped by. Bear, who guarded the third door, was also sleeping. At the fourth door, Coyote found the guardian of the fire asleep. Slipping through into the room of Sussistinnako he found him also sleeping.

Coyote quickly lit the cedar brand which was attached to his tail and hurried out. Spider awoke, just enough to know someone was leaving the room.

"Who is there?" he cried. Then he called, "Someone has been here." But before he could awaken the sleeping Bear and Cougar and Snake, Coyote had almost reached the upper world.

~⁓

Dog Swims to Steal Fire

– Motu –

Southern Coast of New Guinea, Melanesia

The ancestors of the present people had no fire, and ate their food raw or cooked it in the sun until one day they perceived smoke rising out at sea. A dog, a snake, a bandicoot, a bird, and a kangaroo all saw this smoke and asked, "Who will go to get fire?" First the snake said that he would make the attempt, but the sea was too rough, and he was compelled to come back. Then the bandicoot went, but he, too, had to return. One after another, all tried but the dog, and all were unsuccessful. Then the dog started and swam and swam until he reached the island whence the smoke rose. There he saw women cooking with fire, and seizing a blazing brand, he ran to the shore and swam safely back with it to the mainland, where he gave it to all the people.

~⁓

The Dog Teaches Man about Fire

– Dyak –

Baram District, Borneo

One day when the man and the dog were in the jungle together, and got drenched by rain, the man noticed that the dog warmed himself by rubbing against a huge creeper (called the Aka Rawa), whereupon the man took a stick and rubbed it rapidly against the Aka Rawa, and to his surprise obtained fire. Later some food was accidentally dropped near the fire, and the man, finding it thus rendered more agreeable to the taste, discovered the art of cooking.

Grandmother Spider Steals the Fire

– Choctaw –

Southeastern United States

*T*he Choctaw People say that when the People first came up out of the ground, People were encased in cocoons, their eyes closed, their limbs folded tightly to their bodies.

And this was true of all People, the Bird People, the Animal People, the Insect People, and the Human People. The Great Spirit took pity on them and sent down someone to unfold their limbs, dry them off, and open their eyes.

But the opened eyes saw nothing, because the world was dark, no sun, no moon, not even any stars. All the People moved around by touch, and if they found something that didn't eat them first, they ate it raw, for they had no fire to cook it.

All the People met in a great powwow, with the Animal and Bird People taking the lead, and the Human People hanging back. The Animal and Bird People decided that life was not good, but cold and miserable. A solution must be found! Someone spoke from the dark, "I have heard that the people in the East have fire."

This caused a stir of wonder, "What could fire be?" There was a general discussion, and it was decided that if, as rumor had it, fire was warm and gave light, they should have it too. Another voice said, "But the people of the East are too greedy to share with us." So it was decided that the Bird and Animal People should steal what they needed, the fire!

But, who should have the honor? Grandmother Spider volunteered, "I can do it! Let me try!" But at the same time, Opossum began to speak. "I, Opossum, am a great chief of the animals. I will go to the East and since I am a great hunter, I will take the fire and hide it in the bushy hair on my tail." It was well know that Opossum had the furriest tail of all the animals, so he was selected.

When Opossum came to the East, he soon found the beautiful, red fire, jealously guarded by the people of the East. But Opossum got closer and closer until he picked up a small piece of burning wood, and stuck it in the hair of his tail, which promptly began to smoke, then flame. The people of the East said, "Look, that Opossum has stolen our fire!"

They took it and put it back where it came from and drove Opossum away. Poor Opossum! Every bit of hair had burned from his tail, and to this day, opossums have no hair at all on their tails.

Once again, the powwow had to find a volunteer chief. Grandmother Spider again said, "Let me go! I can do it!" But this time a bird was elected, Buzzard. Buzzard was very proud. "I can succeed where Opossum has failed. I will fly to the East on my great wings, then hide the stolen fire in the beautiful long feathers on my head."

The birds and animals still did not understand the nature of fire.

So Buzzard flew to the East on his powerful wings, swooped past those defending the fire, picked up a small piece of burning ember, and hid it in his head feathers. Buzzard's head began to smoke and flame even faster! The people of the East said, "Look! Buzzard has stolen the fire!" And they took it and put it back where it came from.

Poor Buzzard! His head was now bare of feathers, red and blistered looking. And to this day, buzzards have naked heads that are bright red and blistered.

The powwow now sent Crow to look the situation over, for Crow was very clever. Crow at that time was pure white, and had the sweetest singing voice of all the birds. But he took so long standing over the fire, trying to find the perfect piece to steal that his white feathers were smoked black. And he breathed so much smoke that when he tried to sing, out came a harsh, "Caw! Caw!"

The Council said, "Opossum has failed. Buzzard and Crow have failed. Who shall we send?"

Tiny Grandmother Spider shouted with all her might, "LET ME TRY IT, PLEASE!" Though the council members thought Grandmother Spider had little chance of success, it was agreed that she should have her turn. Grandmother Spider looked then like she looks now, she had a small torso suspended by two sets of legs that turned the other way. She walked on all of her wonderful legs toward a stream where she had found clay.

With those legs, she made a tiny clay container and a lid that fit perfectly with a tiny notch for air in the corner of the lid. Then she put the container on her back, spun a web all the way to the East, and walked tiptoe until she came to the fire. She was so small, the people from the East took no notice. She took a tiny piece of fire, put it in the container, and covered it with the lid. Then she walked back on tiptoe along the web until she came to the People. Since they couldn't see any fire, they said, "Grandmother Spider has failed."

"Oh no," she said, "I have the fire!" She lifted the pot from her back, and the lid from the pot, and the fire flamed up into its friend, the air. All the Birds and Animal People began to decide who would get this wonderful warmth. Bear said, "I'll take it!" but then he burned his paws on it and decided fire was not for animals, for look what happened to Opossum!

The Birds wanted no part of it, as Buzzard and Crow were still nursing their wounds. The insects thought it was pretty, but they, too, stayed far away from the fire.

Then a small voice said, "We will take it, if Grandmother Spider will help." The timid humans, whom none of the animals or birds thought much of, were volunteering!

So Grandmother Spider taught the Human People how to feed the fire sticks and wood to keep it from dying, how to keep the fire safe in a circle of stone so it couldn't escape and hurt them or their homes. While she was at it, she taught the humans about pottery made of clay and fire, and about weaving and spinning, at which Grandmother Spider was an expert.

The Choctaw remember. They made a beautiful design to decorate their homes, a picture of Grandmother Spider, two sets of legs up, two down, with a fire symbol on her back. This is so their children never forget to honor Grandmother Spider, Fire Bringer.

~⌒~

The Hungry Fire

– Alabama-Coushatta –

Southeastern United States

*O*nce a Fire was almost burned out and was making the hissing sound then usually heard. When a man asked it what it wanted, the Fire said, "Something to eat." "What do you want to eat?" "I want to eat wood." So the man picked up some dead wood and piled a quantity of it on the Fire. The Fire grew bigger and bigger, and the man kept piling on more and more wood, until the Fire cried to the man to hold all of the animals back so that they would not be burned.

~⌒~

Bears Lose Fire to Humans

– Koasati (Coushatta-Alabama) –

Southeastern United States

*B*ears formerly owned the Fire and they always took it about with them. One time they set it on the ground and went on farther eating acorns. The Fire nearly went out and called aloud. It was almost extinguished. "Feed me," it said.

Then some human beings saw it. They found a stick north of the fire and placed it over the flames. They found another stick west and placed it on top. They found a stick south of the fire and finally east of the fire, and placed them on top in order. The Fire blazed up.

When the bears came to get their Fire, it said, "I don't know you anymore." They did not get it back and so it belongs to human beings.

Wolverine Freezes to Death

– Lnu'k (Mi'kmaq) –

Northern Maine

*O*ne day Wolverine visited his older brother Bear, who was very glad to see him, and at once put the pot on the fire to cook him something. After the food was cooked and they had eaten it, Bear said to his younger brother Wolverine, "How would you make a fire if you did not have any flint and steel?" Wolverine acknowledged that he would be helpless without flint and steel. "Now I will teach you," said Bear, "how to make a fire, when you do not have any flint and steel." Having said this, Bear went out and got some maple bark, which he put in a little pile, and then jumped over it. As soon as he jumped over it, it burst into a flame. Then he said to his younger brother, "Now I give you power to make a fire."

Wolverine was very happy and was in a hurry to get away and try his power. As soon as he got out of the house, he started to run. He continued running until he got to a place where he could no longer see Bear. Then he collected some maple bark and made a little pile of it and jumped over it. When it broke into a blaze, he was very much pleased. He took out his flint and steel and threw them away, saying, "These are no longer of any use."

Wolverine had no use for the fire he made; he only made it to try his power. So he went on, but he had hardly gotten out of sight of his first fire, when he decided to make a new fire. After that he made fires more frequently until at last he made them every ten steps; but finally his power gave out, for he had used it all up. When he next collected a pile of maple bark and jumped over it, it did not burst into flame. By that time it had grown dark and was very cold, and he was indeed in need of a fire. Then truly he jumped, but no success crowned his efforts. He had thrown away his flint and steel and was very much frightened, for it was very cold. He kept on jumping, but it grew so cold that he froze to death while he was jumping. He lay there until spring, when he thawed out. He was lying there dead, when his younger brother, Raccoon, came along and saw him.

Raccoon went over and tried to wake him up, saying, "Older brother, get up, you are over-sleeping, it is very late." Then the Wolverine rubbed his eyes, got up and said, "Younger brother, I overslept. I would have lain there forever, if you had not come by and awakened me." He would have rotted there, but as it was, he got his strength back and was as strong as ever.

~

Rabbit Steals Fire

– Hitchiti –

Chattahoochee River Watershed, Georgia

Rabbit was traveling about during a festal season. On those days fire was taboo until they built a fire in the ceremonial grounds. Rabbit knew when there was to be a dance at the busk ground and thought, "I will run away with some fire." He considered the matter and decided how he would do it. He had his head rubbed with pine tar so as to make his hair stand up. Then he set out. When he arrived at the busk ground a great number of people were gathered there. While Rabbit was sitting about and the people were dancing, they said that he must lead and he agreed. "Now, lead," they said, and he got up and danced ahead of them around the place where the fire was. As he went many people followed him and Rabbit started the song. He was dancing along, the rest following him. While they were dancing very hard Rabbit ran round near the fire and bent his head as if he were going to take hold of it. They said, "When he is leading Rabbit always acts that way." He kept on acting that way and circled about as he did so. Presently he poked his head into the fire and ran off with his head ablaze, while the people shouted, "Hulloa, catch him or throw him down."

They shouted at Rabbit as he ran away, and they chased him, but he disappeared. Then they made it rain and on the fourth day they said the rain must have put the fire out. So it stopped and the sun shone and the weather was fine. But Rabbit had built a fire in a hollow tree and stayed there while it rained, and when the sun shone he came out and set out fires. Rain came on again and put the fires all out but he again built a fire inside of a hollow tree. When the sun shone he would set out fires and then rains would come and put them out, but they could not stop them entirely. People took fire and ran off with it. Rains kept on putting the fires out at intervals but when they stopped all the people distributed it again, and when the rain stopped fire was established there for good. This is the way it is told. Therefore, they say that Rabbit distributed the fire to all people.

Ti'iti'i Steals Fire from Mafuie

– Polynesian –

Samoa

The home of Mafuie the earthquake god was in the land of perpetual fire. Ti'iti'i's father Talanga was also a resident of the underworld and a great friend of the earthquake god.

Ti'iti'i watched his father as he left his home in the upper-world. Talanga approached a perpendicular wall of rock, said some prayer or incantation, and passed through a door which immediately closed after him.

Ti'iti'i went to the rock, but could not find the way through. He determined to conceal himself the next time so near that he could hear his father's words.

After some days he was able to catch all the words uttered by his father as he knocked on the stone door: "O rock! Divide. I am Talanga. I come to work on my land given by Mafuie."

Ti'iti'i went to the perpendicular wall and imitating his father's voice called for a rock to open. Down through a cave he passed until he found his father working in the underworld.

The astonished father, learning how his son came, bade him keep very quiet and work lest he arouse the anger of Mafuie. So for a time the boy labored obediently by his father's side.

In a little while the boy saw smoke and asked what it was. The father told him that it was the smoke from the fire of Mafuie, and explained what fire would do.

The boy determined to get some fire. He went to the place from which the smoke arose and there found the god, and asked him for fire. Mafuie gave him fire to carry to his father. The boy quickly had an oven prepared and the fire placed in it to cook some of the taro they had been cultivating. Just as everything was ready an earthquake god came up and blew the fire out and scattered the stones of the oven.

Then Ti'iti'i was angry and began to talk to Mafuie. The god attacked the boy, intending to punish him severely for daring to rebel against the destruction of the fire.

A severe battle ensued. At last Ti'iti'i seized one of the arms of Mafuie and broke it off. He caught the other arm and began to twist and bend it.

Mafuie begged the boy to spare him. His right arm was gone. How could he govern the earthquakes if his left arm were torn off also? It was his duty to hold Samoa level and not permit too many earthquakes. It would be hard to do that even with one arm, but it would be impossible if both arms were gone.

Ti'iti'i listened to the plea and demanded a reward if he should spare the left arm. Mafuie offered Ti'iti'i one hundred wives. The boy did not want them.

Then the god offered to teach him the secret of finding fire to take to the upper-world. The boy agreed to accept the fire secret, and thus learned that the gods in making the earth had concealed fire in various trees for men to discover in their own good time, and that this fire could be brought out by rubbing pieces of wood together.

Rabbit Steals Fire from the East

– Muscogee (Creek) –

Southeastern United States

*A*ll the people came together and said: "How shall we obtain fire?" It was agreed that Rabbit should try to obtain fire for the people.

He went across the great water to the east. He was received gladly, and a great dance was arranged. Then Rabbit entered the dancing circle, gaily dressed, and wearing a peculiar cap on his head into which he had stuck four sticks of rosin.

As the people danced they approached nearer and nearer the sacred fire in the center of the circle. The Rabbit also danced nearer and nearer the fire. The dancers began to bow to the sacred fire, lower and lower. Rabbit also bowed to the fire, lower and lower. Suddenly, as he bowed very low, the sticks of rosin caught fire and his head was a blaze of flame.

The people were amazed at the impious stranger who had dared to touch the sacred fire. They ran at him in anger, and away ran Rabbit, the people pursuing him. He ran to the great water and plunged in, while the people stopped on the shore.

Rabbit swam across the great water, with the flames blazing from his cap. He returned to his people, who thus obtained fire from the east.

Goorda Learns to Share Fire

– Aborigine –

Australia

*G*oorda, the fire spirit, lived alone among the stars of the Southern Cross. He was a great hunter who traveled between three different campfires, known as the Pointers. When he was lonely, Goorda wanted his neighbors to visit.

"Would you like to come to my camp?" he asked. "I will give you as much as you can eat, and we can share songs and stories."

Goorda prepared his camp. He watched and hoped that the visitors would come, but no one arrived. Down below, on Earth, Goorda saw that people lived together. They helped each other hunt and gather roots and nuts. Children swam and played games with their friends. Adults told stories when they gathered in the evening. "Come," Goorda called down to Earth, "there is much game where I live. You may hunt here." But Earth's people did not come.

One night, Goorda saw the Earth people eating raw meat from the kangaroo and the goanna lizard. He watched them sit close together in order to keep warm as they ate. "Fire, that is what the Earth people do not have," he said to himself. "But fire is something I have enough of for everyone. I will bring it to the Earth people."

"Please take me with you!" Goorda cried to one of the falling stars as it streaked toward Earth. But it did not stop. After he banked his fires so they would not go out, Goorda followed the path of the falling star and shot toward Earth in a bright flash.

He headed for the shore opposite where a group of people were gathered along the banks of the Gainmaui River by Caledon Bay.

"I come bearing a gift for you," said Goorda. When his feet touched ground, however, the brush began to crackle with flames. Soon, a blaze was roaring along the riverbank. Across the river, the people picked up their spears and watched, their faces kissed by the red hues of the dancing flames.

When the heat from the fire became so great they could feel its pain, they screamed and fled from Goorda.

"Stay! Do not go!" Goorda cried, "I will not hurt you." But flames spread everywhere he walked. Those people who jumped into their canoes and paddled away from shore survived. A goanna lizard burrowed into the ground and closed its doorway with soil. It, too, escaped the smoke and flames. Nearby, the spider, Garwuli, survived by taking refuge in the deep fissure of a broken rock.

A cloud of honeybees rose into the air and fled to the shelter of a hollow tree a great distance away. The swifts circled overhead and caught insects that fled upward from the flames.

Crocodiles, barramundi fish and other water creatures swam away from the hot, steaming waters along the riverbank. All of these escaped Goorda's fire.

But Goorda, not understanding, tried to get close to people so he could visit with them. Some people were burned in their huts.

Others fled in terror. By the end of the day, Goorda looked out over a silent, black, smoldering land. Overhead, a red-winged parrot called and circled in search of a place to land, then flew away searching for safety until it sank down below the horizon in the distance.

Exhausted, Goorda sat down to rest. He found a kangaroo that had been burned and began to eat. "This will not do," he said.

"I will have to appear to the Earth people in a different way, or they will always flee from me." Goorda changed to the form of a person and painted a diamond on his breast. "This will be my symbol," he said. "Now I am ready to greet people and teach them how to use fire in a good way."

When the sun rose over the blackened land Goorda had created, a curious group of hunters came wandering. Goorda saw them and stood up. "Hello, my friends," he said. Flames began to shoot from the end of the stick he held in his hand. "Do not fear," said Goorda. "Fire can be of use to you. It can cook your food." Saying that, Goorda picked up a piece of blackened kangaroo, took a bite and chewed.

"See, it tastes much better this way."

Goorda handed the meat to the men and each carefully took a bite as they passed it around. "Indeed, it is very good," they agreed. They searched around for more meat and ate their fill. As they were eating, Goorda showed them the ways of fire. He demonstrated how to create fire by holding a stick between the hands and turning it quickly while the end sits in a small hole bored into another piece of wood. In a short time, the light of an ember appeared where the two sticks met.

"Be careful with flame," said Goorda. "Clear the brush away from your campfires and watch the flames closely so they do not spread.

"When the leaves and branches are dry, you can light a fire to herd the animals so hunting will be easier. Later, when the rains come and the Moon Man has once gone and returned, those trees and bushes will sprout new leaves and twigs for the wallaby and kangaroo to eat."

The hunters learned everything Goorda had to teach, then they thanked him and walked back to tell their families.

Satisfied with his visit, Goorda again took on the form of the spirit. He streaked up through the icy blackness toward his home in the Southern Cross. The flames of his fires continue to dance and flicker in the night skies.

～⚬

How Coyote Stole Fire

– Shasta –

Northwestern California, Southwestern Oregon

ℒong ago, when man was newly come into the world, there were days when he was the happiest creature of all. Those were the days when spring brushed across the willow tails, or when his children ripened with the blueberries in the sun of summer, or when the goldenrod bloomed in the autumn haze.

But always the mists of autumn evenings grew more chilled, and the sun's strokes grew shorter. Then man saw winter moving near, and he became fearful and unhappy. He was afraid for his children, and for the grandfathers and grandmothers who carried in their heads the sacred tales of the tribe. Many of these, young and old, would die in the long, ice bitter months of winter.

Coyote, like the rest of the People, had no need for fire. So he seldom concerned himself with it, until one spring day when he was passing a human village. There the women were singing a song of mourning for the babies and the old ones who had died in the winter. Their voices moaned like the west wind through a buffalo skull, prickling the hairs on Coyote's neck.

"Feel how the sun is now warm on our backs," one of the men was saying. "Feel how it warms the earth and makes these stones hot to the touch. If only we could have had a small piece of the sun in our tipis during the winter."

Coyote, overhearing this, felt sorry for the men and women. He also felt that there was something he could do to help them. He knew of a faraway mountaintop where the three Fire Beings lived. These Beings kept fire to themselves, guarding it carefully for fear that man might somehow acquire it and become as strong as they. Coyote saw that he could do a good turn for man at the expense of these selfish Fire Beings.

So Coyote went to the mountain of the Fire Beings and crept to its top, to watch the way that the Beings guarded their fire. As he came near, the Beings leaped to their feet and gazed searchingly round their camp. Their eyes glinted like bloodstones, and their hands were clawed like the talons of the great black vulture.

"What's that? What's that I hear?" hissed one of the Beings.

"A thief, skulking in the bushes!" screeched another.

The third looked more closely, and saw Coyote. But he had gone to the mountain top on all fours, so the Being thought she saw only an ordinary coyote slinking among the trees.

"It is no one, it is nothing!" she cried, and the other two looked where she pointed and also saw only a gray coyote. They sat down again by their fire and paid Coyote no more attention.

So he watched all day and night as the Fire Beings guarded their fire. He saw how they fed it pine cones and dry branches from the sycamore trees. He saw how they stamped furiously on runaway rivulets of flame that sometimes nibbled outwards on edges of dry grass. He saw also how, at night, the Beings took turns to sit by the fire. Two would sleep while one was on guard; and at certain times the Being by the fire would get up and go into their tipi, and another would come out to sit by the fire.

Coyote saw that the Beings were always jealously watchful of their fire except during one part of the day. That was in the earliest morning, when the first winds of dawn arose on the mountains. Then the Being by the fire would hurry, shivering, into the tipi calling, "Sister, sister, go out and watch the fire." But the next Being would always be slow to go out for her turn, her head spinning with sleep and the thin dreams of dawn.

Coyote, seeing all this, went down the mountain and spoke to some of his friends among the People. He told them of hairless man, fearing the cold and death of winter. And he told them of the Fire Beings, and the warmth and brightness of the flame. They all agreed that man should have fire, and they all promised to help Coyote's undertaking.

Then Coyote sped again to the mountain top. Again the Fire Beings leaped up when he came close, and one cried out, "What's that? A thief, a thief!"

But again the others looked closely, and saw only a gray coyote hunting among the bushes. So they sat down again and paid him no more attention.

Coyote waited through the day, and watched as night fell and two of the Beings went off to the tipi to sleep. He watched as they changed over at certain times all the night long, until at last the dawn winds rose.

Then the Being on guard called, "Sister, sister, get up and watch the fire." And the Being whose turn it was climbed slow and sleepy from her bed, saying, "Yes, yes, I am coming. Do not shout so."

But before she could come out of the tipi, Coyote lunged from the bushes, snatched up a glowing portion of fire, and sprang away down the mountainside.

Screaming, the Fire Beings flew after him. Swift as Coyote ran, they caught up with him, and one of them reached out a clutching hand. Her fingers touched only the tip of the tail, but the touch was enough to turn the hairs white, and coyote tail tips are white still. Coyote shouted, and flung the fire away from him. But the others of the People had gathered at the mountain's foot, in case they were needed. Squirrel saw the fire falling, and caught it, putting it on her back and fleeing away through the tree-tops. The fire scorched her back so painfully that her tail curled up and back, as squirrels' tails still do today.

The Fire Beings then pursued Squirrel, who threw the fire to Chipmunk. Chattering with fear, Chipmunk stood still as if rooted until the Beings were almost upon her. Then, as she turned to run, one Being clawed at her, tearing down the length of her back and leaving three stripes that are to be seen on chipmunks' backs even today. Chipmunk threw the fire to Frog, and the Beings turned towards him. One of the Beings grasped his tail, but Frog gave a mighty leap and tore himself free, leaving his tail behind in the Being's hand—which is why frogs have had no tails ever since.

As the Beings came after him again, Frog flung the fire on to Wood. And Wood swallowed it.

The Fire Beings gathered round, but they did not know how to get the fire out of Wood. They promised it gifts, sang to it and shouted at it. They twisted it and struck it and tore it with their knives. But Wood did not give up the fire. In the end, defeated, the Beings went back to their mountaintop and left the People alone.

But Coyote knew how to get fire out of Wood. And he went to the village of men and showed them how. He showed them the trick of rubbing two dry sticks together, and the trick of spinning a sharpened stick in a hole made in another piece of wood. So man was from then on warm and safe through the killing cold of winter.

Arámfè Sends Fire to Men

– Yorubas –

Southwestern Nigeria

...Odudúwa sent
Ífa, the Messenger, to his old sire
To crave the Sun and the warm flame that lit
The torch of Heaven's Evening and the dance. . .
A deep compassion moved thunderous Arámfè,
The Father of the Gods, and he sent down
The vulture with red fire upon his head
For men; and, by the Gods' command, the bird
Still wears no plumage where those embers burned him—
A mark of honour for remembrance.

Acknowledgments

𝒲riting this book has empowered me to walk in harmony better, especially with fire, throughout my life than I had previously. However, I didn't take this walk alone. I am grateful to the people who helped me along the way.

I thank my wife, daughters and parents for their support throughout my writing process. Our shared experiences clearly influenced components of these stories.

Several people generously contributed their insights and thus strengthened my arguments: Brad Jackson, Deanna Harrington, Michelle Johnson, Fred James, Assistant Imam Fudail Hassan, Karen Hajek, Cheryl Poage and Lisa Miklas. I also thank Warren Yahr for sharing permission to use his story after our telephone conversation and Glenn Welker for sharing the story of Grandmother Spider so freely. Jennifer Ott and Teresa Burden read early versions of this manuscript while Susie Keppers, Hilary Bilbrey, Deanna Harrington and Maria Bostian read the final draft; I thank these mentors and colleagues for their time and expertise. If there are any mistakes, errors, shortcomings or misinterpretations, please blame the messenger; I own them.

I appreciate the immense professional and personal support I receive from my coworkers at South Metro Fire Rescue Authority, peers in the community risk reduction field and teachers in Douglas County and Cherry Creek School Districts. I also value my time with Evergreen Fire Rescue, where I formally entered the risk reduction world, and I treasure my time as a firefighter with Idaho Springs Fire Department, Clear Creek Fire Authority and the Clear Creek Sheriff's Office Marmot Wildfire Crew.

I conducted much of my research online, but went the extra mile to find original documents. The historian in me smiles knowing that so many older books have been digitized by the Google folk. The efforts of countless, nameless Googlers provided resources that otherwise would have been beyond my reach.

My background in environmental history influences how I analyze, act on and educate others about risk. My mentors at the University of Montana and University of Puget Sound, including Dan Flores, David Emmons, Harry Fritz, Drew Isenberg, Nancy Bristow, Bill Breitenbach and Petra Goedde taught me how to think and write about history. They also demonstrated the importance of learning and sharing.

When I was close to finishing this manuscript, I considered self-publishing, but I wanted more for *Ancient Fire, Modern Fire*. I found amazing collaborators in LaVonne and Rex Ewing of PixyJack Press. They helped sharpen my writing and strengthen my arguments by looking beyond Colorado. They also helped me maneuver the process of publishing. I am profoundly grateful for their assistance, professionalism, support and growing friendship.

Survey Tools for Youth Fire Misuse

*T*he Juvenile Firesetter Child and Family Risk Surveys offer an accurate means to assess the risk of future firesetting among children. The roots of these surveys stretch back to Dr. Kenneth Fineman's original work for the US Fire Administration in the 1970s. He and a group of Coloradans created these surveys in the 1990s to empower fire departments to conduct assessments more quickly in a non-clinical setting. They consist of two sections: one gathers information about the child who set the fire while the other gathers information about the child's home and family from a parent or caregiver. The survey process also protects the family's privacy and the agency conducting the interview.

As I finish this manuscript, a pair of 10-year-old kids in Taylorsville, Utah, have been charged with aggravated arson. They were playing with lighters in a tree. The fire spread into the attic of an adjacent apartment building. The fire destroyed eight apartments and damaged others, displacing twenty people. There were no injuries among the kids, residents or firefighters, fortunately. An 8-year-old child was not charged, but was present for the fire misuse.

These children—*were they kids just being kids?*—and thousands of others across the country destroy $200 million in property annually, according to the NFPA. Their fires also cause hundreds of injuries and scores of deaths. Enough is enough. These surveys help families and agencies prevent future fires when they identify children at risk and when the families acquire the help they need.

If you are using these tools because your child has set a fire, I highly encourage you to contact your local fire department, children's hospital or mental health agency so you and your family can take the next step toward addressing the risk. Community risk reduction depends on communities working together.

FAMILY FIRE RISK SURVEY

Administered to parents / guardians over the phone or in person.

The Family Risk Survey is designed to be given to parents who have concerns about their child's fire play or firesetting behavior, or whose child has set a fire which has come to the attention of a fire department, policy agency or other community agency. It is intended for use only as a preliminary screening tool and should be used with the Child Risk Survey to assess the child's suitability for fire intervention education or mental health referral.

Incident & Demographic Information

Data to be gathered includes information about the incident, the child's name, address, school, grade level, name of parent/guardian, and referral source (such as caregiver, parent, school, law enforcement, mental health provider, fire service, juvenile justice).

Questions for the Family

Ask the question as written, check the response, place the appropriate constant weight in the score column, and add the scores to determine the total family risk score. Please substitute the child's name in questions 1-5. Score only one response per question, using the one with the highest risk value.

	CONSTANT	SCORE

1. If you had to describe (child's name) curiosity about fire, would you say it was absent, mild, moderate, or extreme?

	CONSTANT	SCORE
Absent	0	_____
Mild	99	_____
Moderate	198	_____
Extreme	297	_____

2. Has (child's name) been diagnosed with any impulse control conditions, such as Attention Deficit Disorder (ADD) or ADD with hyperactivity (ADHD)?

	CONSTANT	SCORE
Yes; Diagnosis:_____	28	_____
No	0	_____

3. Has (child's name) been in trouble outside of school for non-fire related behavior?
    Yes; What? _____ 90   _____
    No   0   _____

4. Has (child's name) ever stolen or shoplifted?
    Yes   14   _____
    No   0   _____
    DK/NA   0   _____

5. Has (child's name) ever beat up or hurt others?
    Yes   14   _____
    No   0   _____
    DK/NA   0   _____

6. Besides this fireplay or firesetting incident, how many other times has your child played with fire, including matches or lighters, or set something on fire?
    1 (current)   84   _____
    2 (current + 1)   168   _____
    4 (current + 2-4)   336   _____
    6 (current + 5)   504   _____

7. Is there an impulsive (sudden urge) quality to your child's firesetting or fireplay?
    Yes   71   _____
    No   0   _____
    DK/NA   0   _____

8. Informational purposes only: Is there a history of emotional, physical, or sexual abuse in the family? ❏ YES ❏ NO
    Who? / Relationship? / Currently in home?

Total Risk Family Score   _____

FAMILY RISK
Little Risk = <429 (Education)
Definite Risk = 429<457 (Referral + Education)
Extreme Risk = >457 (Referral)

CHILD RISK SURVEY

Administered to the child, in person, and separate from their parents, only after the parents or guardians have provided written informed consent for the child's participation

The Child Risk Survey is designed to be given to children who have played with fire or who have set a fire which has come to the attention of a fire department, police agency or the community agencies. It is intended for use only as a preliminary screening tool and should be used with the Family Risk Survey to assess the child's suitability for fire intervention education or mental health referral.

Incident & Demographic Information

If not already provided in the Family Risk Survey: information about the incident, child's name, address, school, grade level, name of parent/guardian, and referral source (such as caregiver, parent, school, law enforcement, mental health provider, fire service, juvenile justice).

Informational Activity for the Child

Have the child draw a picture of the fire or fireplay incident and/or write a paragraph describing why they are in your office today while you are conducting the Family Survey with the parents.

Development of Rapport

The purpose of this section is to make the child comfortable with you. The more at ease you can make him, the greater the likelihood that he will answer all of your questions. If the following questions aren't enough, add your own. Questions or language can be modified in the Development of Rapport section only, **all other questions should be asked as written**.

[Introduce yourself]
1. What's your name?
2. How old are you?
3. What school do you go to?

What grade are you in?

Do you like your school?

Are there nice/okay teachers at your school?

4. What classes/subjects do you like/not like?
5. What do you do for fun? Do you have hobbies?
6. Who's your best friend?
7. What do you like to play/do with your friend?
8. What do you watch on TV and/or what videos do you watch?
9. What is your favorite person/show on TV?
10. What is your favorite video/computer game?
11. What do you like about that game? [Is there extreme interest in violence or fire?]

When rapport is established, determine level of understanding if the child is under 7 or appears to have problems communicating.

Determine Level of Understanding (Under Age 7)

It is often difficult to determine if a young child really understands you. (This section may be skipped if you are interviewing an older child). There may be an age barrier, a language barrier, a learning problem, or sub-normal intelligence. It is fruitless to go through an entire interview unless you are first assured that the child has enough understanding to complete the interview. There are several ways to gauge whether you are on the same "wave length" as the child. The following are suggested ways to do so:

a. Obtain information from rapport section above:
By paying close attention to the manner in which a young child responds to the 11 questions above, you can estimate whether he can understand and respond to the other questions in this instrument.

b. Using crayons/paper as a tool:
You can ask the child to draw pictures of common objects, his favorite toys, houses, trees, and people. Then ask him to describe what he has drawn. Clear explanations of his drawings and the action taking place in some of those drawings will tell you something about the child's vocabulary and his ability to understand.

c. **Using toys and games:**

Have toys of the appropriate developmental level of the child available. Engage the child in a game with the toys or allow the child free play with the toys. After a while ask the child about the toys and the game he is playing. Inquire about the rules, the purpose, etc. Estimate the child's vocabulary in terms of his ability to complete the interview.

d. **Using puppets:**

Have hand puppets available. Allow the child to set the interaction, with the child playing all parts or with you playing some of the parts. Quiet children can become quite verbal with this approach. Focus on the child's ability to understand your questions during the puppet play and determine if this level of communication is sufficient for continued interviewing.

If you are satisfied that the child has adequate understanding, proceed with the questions on the next page.

Comparison of the Order of Events

For children age nine and older, have the child describe their involvement in the incident from some point in time prior to the incident to some point in time after the incident. At the end of the interview ask the child to repeat this description in reverse order.

The average child who is nine years old should be able to explain the incident in reverse order. If the events in reverse differs significantly from the events as they actually happened, the child may be trying to hide aspects of the incident.

Developed by Kenneth R. Fineman Ph.D., and is reprinted from Comprehensive Fire Risk Assessment as published in the Colorado Juvenile Firesetter Prevention Program: Training Seminar Vol. I.

Questions for the Child

Ask the question as written, check the response, place the appropriate constant weight in the score column, and add the scores to determine the Total Child Risk Score. Score only one response per question, using the one with the highest risk value.

	CONSTANT	SCORE

1. Do you have any brothers or sisters?

Yes	0	_____
No (If no, skip to #3)	0	_____

2. How well do you get along with them?

Always get along	28	_____
Usually get along	56	_____
Sometimes get along	84	_____
Don't get along very often	112	_____
Never get along	140	_____

3. How well do you get along with your mother?

Always get along	10.5	_____
Usually get along	21	_____
Sometimes get along	31.5	_____
Don't get along very often	42	_____
Never get along	52.5	_____

4. Do you fight or argue with your mother?

Never	10.5	_____
Rarely	21	_____
Sometimes	31.5	_____
Usually	42	_____
Always	52.5	_____

5. Do you see your father as much as you'd like?

Yes	0	_____
No	60	_____
Too much	60	_____

6. When you are asked to do something, do you usually do it?
   Yes   0   _____
   No   17.5   _____

7. Do you lie a lot?
   Yes   17.5   _____
   No   0   _____

8. What happens at home when you get in trouble?
   Grounded   0   _____
   Physical punishment   0   _____
   Talked/Lectured   0   _____
   Sought outside help   0   _____
   Abused**   0   _____
   Other/nothing   0   _____
   Yelled at   32   _____

9. Has there been an ongoing (chronic) crisis or problem in your life or in your family?
   Yes (what?)   62   _____
   No   0   _____

10. Besides this fireplay or firesetting incident, how many other times have you played with fire, including matches or lighters, or set something on fire?
    1 (current)   32   _____
    2 (current + 1)   64   _____
    4 (current + 2-4)   128   _____
    6 (current + 5)   192   _____

11. What did you do after the fire started?
    Put it out | Called for help   0,0   _____
    Ran away | Didn't try to run   0,0   _____
    Panicked | Tried to extinguish   0,0   _____
    Other | Didn't try to extinguish   0,0   _____
    Stayed & watched   40   _____

12. Did you intend to play with fire or set the fire,
    that is, did you play with or set the fire on purpose?
    Yes   187   _____
    No   0   _____
    *If surveyor has evidence of intent, the surveyor
    may override the youth's denial*

13. Where did you set the fire?
    Structure involved as a target
    or a location   47   _____
    Other: 0   _____

14. Do you like to look at fire for long periods of time?
    Yes   250   _____
    No   0   _____

15. How did you get the ignition source (match/lighter/other) used
    in the fire/fireplay? [INFO ONLY]:

Total Child Risk Score   _____

** *If there are indications of abuse or neglect, consult with social services or law enforcement immediately.*

CHILD RISK
Little Risk = <511 (Education)
Definite Risk = 511<540 (Referral + Education)
Extreme Risk = >540 (Referral)

Source: Moynihan, Flesher, and Colorado Juvenile Firesetter Prevention Program Staff; 06/29/98

Online Resources

I reached out to colleagues in community risk reduction to compile this list of resources. Thank you to Eric Gleason, Lenore Corey, Cindy Kettering, Karla Klas and Nancy Kidd for sharing their favorite online resources. While some of these websites only target certain audiences and certain topics, each of them is valuable as a tool for creating better relationships with fire.

GENERAL

www.firesafetyeducators.org
The Fire & Life Safety Educators of Colorado is a network of community risk reduction professionals based on Colorado. The website contains lesson plans and contacts for fire departments in Colorado.

www.fmac-co.org
The Fire Marshal's Association of Colorado is committed to preventing and mitigating fire through education, enforcement, engineering and investigation. This website and the network of professionals it represents are experts in interpreting, developing and applying fire codes.

washingtonfirechiefs.com/Sections/PublicFireEducators
Washington Public Fire Educators has a robust website with many links to even more websites that discuss fire prevention and fire safety.

HOME FIRE SAFETY

www.campusfiresafety.org
This website is a resource for families with children in college. On-campus and off-campus living have their risks, but those risks can be mitigated easily when those college students boost their awareness and take responsibility.

www.childrensvillagehagerstown.org
There are many fire safety villages in the US and Canada. I've been to and was impressed by the one in Hagerstown, MD.

www.firefacts.org
A safe house is a happy home. That's the premise of this website that contains resources for school-aged children, parents and teachers. A related website is **www.firesafekids.org**.

www.fireproofchildren.com
This website is the home of "play safe! be safe!" and other materials created with assistance from the BIC Corporation.

www.homefiredrill.org
You'll find all the tools you need for a home fire drill at this website.

www.homefiresprinkler.org
Source for information on residential sprinkler systems. Materials are designed specifically for consumers, the fire service, home builders and other professionals.

www.njfiresafety.com
Fire is black, fire is hot and fire is plenty of other things. This website, hosted by the Saint Barnabas Medical Center in New Jersey, shatters the myths to give 5–7 year olds (and other ages) a better handle on fire's dangers.

www.oregon.gov/osp/SFM
The Oregon Office of the State Fire Marshal has a curriculum for middle school kids called "It's Up To You" that educates on the realities of fire and busts myths propagated by media in order to prevent youth fire incidents. It addresses fire science, prevention, and personal accountability.

www.safekids.org/fire
Safe Kids Worldwide targets children for its messaging. This page is the gateway to its resources for home fire prevention. It includes resources for children with special needs.

www.sosfires.com
This website has resources for youth (and their families) who misuse fire. It contains information for parents, teachers and life safety professionals interested in creating a program for assessing and helping youth who misuse fire.

www.sparky.org
This interactive website is the official home of Sparky the Fire Dog, the children's mascot for the National Fire Protection Association. There are activities, mobile apps, cartoons, games and other goodies here. Grown-ups can find a wealth of information at **www.nfpa.org**.

www.traumaburn.org
The staff of the University of Michigan's Trauma Burn Center developed Sean's Story, an evidence-based program that is among the best in the nation for preventing and responding to fire misuse by children.

www.usfa.fema.gov/prevention/ outreach/children.html
The US Fire Administration also maintains websites for children and adults. The no-nonsense adult information is at **www.usfa.fema.gov**.

WILDFIRE

www.csfs.colostate.edu/pages/ wildfire

Most state forestry agencies will have wildfire mitigation topics. I'm partial to Colorado, of course.

www.firewise.org
This website is the home of FireWise Communities.

www.plt.org
Project Learning Tree is a great program. Its scope is greater than fire science, but it does have multiple lessons regarding fire science and fire ecology.

ppwpp.org
The Pikes Peak Wildfire Prevention Partners is a non-profit network of wildland fire mitigation experts. They primarily serve the residents of Douglas, Teller and El Paso counties in Colorado, but their resources are applicable to a much greater audience.

www.SurvivingWildfire.com
Surviving Wildfire is a useful handbook for homeowners living in the WUI. It will help them get prepared, stay alive, and rebuild their lives if disaster strikes.

BURN PREVENTION

www.ameriburn.org/ preventionEdRes.php
A gateway for support for individuals and their families who experience a burn injury. Also contains great information on preventing burns among all ages.

www.burnprevention.org
Based in eastern Pennsylvania, this organization has programs, products and services available to prevent burn injuries for many age groups.

www.childrenscolorado.org/ departments/surgery/burn
Children's Hospital of Colorado

AGENCY WEBSITES

The following agencies are among thousands across the continent that include solid prevention and mitigation resources on their websites. Most states have a state fire marshal's office. That's a great place to start your research, too. I included the following agencies here because I know the life safety professionals who maintain the materials on the sites and deliver programming in their communities.

www.abbotsford.ca/public_ safety/fire_rescue_service
Abbotsford Fire Rescue Service (BC)

www.cityofalbany.net/ departments/fire
Albany Fire Department (OR)

www.arvadafire.com
Arvada Fire Protection District (CO)

www.aspenfire.com
Aspen Fire Protection District (CO)

www.brfd.org
Boulder Rural Fire Protection District (CO)

www.fire.ca.gov
CALFIRE (CA)

www.friscotexas.gov/ departments/fire
Frisco Fire Department (TX)

www.kentfirerfa.org
Kent Fire Department Regional Fire Authority (WA)

www.ci.loveland.co.us
Loveland Fire Rescue Authority (CO)

www.nltfpd.net
North Lake Tahoe Fire Protection District (NV)

www.plano.gov/215/Fire-Rescue
Plano Fire Department (TX)

www.poudre-fire.org
Poudre Fire Authority (CO)

www.poulsbofire.org
Poulsbo Fire Department (WA)

www.firedistrict1.org
Snohomish County Fire District 1 (WA)

www.springsgov.com
Colorado Springs Fire Dept. (CO)

www.southmetro.org
South Metro Fire Rescue Authority (CO)

www.westpierce.org
West Pierce Fire & Rescue (WA)

www.woodsidefire.org
Woodside Fire Protection District (CA)

Research Notes

Dedication

My twin daughters, then age 6, sang this poem to me one summer night as we prepared a backyard fire for s'mores in the summer of 2014.

Prologue

Most of the information from the Buckweed Fire came from "California Fire Siege 2007: An Overview" published jointly by the California Department of Forestry and Fire Protection (CALFIRE), U.S. Forest Service and California's Office of Emergency Management.

Colleen Morton Busch, *Fire Monks: Zen Mind Meets Wildfire* (New York: Penguin Books, 2012), 54.

1 – Fire, Our Friend and Foe

The introductory quotations are from A.L. Kroeber, *Indian Myths of South Central California,* University of California Publications in American Archaeology and Ethnology Series, ed. Frederic Ward Putnam, vol. 4 (Berkeley: The University Press, 1907), 219; Stephen Pyne, quoted in *Fire Monks: Zen Mind Meets Wildfire* by Colleen Morton Busch (New York: Penguin Books, 2012), i.

"Indians came in..." is from Elliott Coues, ed., *The Manuscript Journals of Alexander Henry, Fur Trader of the Northwest Company, and of David Thompson, Official Geographer and Explorer of the Same Company, 1799-1814,* Vol. I (New York: Francis P. Harper, 1897), 159.

The complete version of August Derleth's poem can be found in his *Collected Poems: 1937-1967,* (New York: The Candlelight Press, 1967), 25-26.

The National Fire Protection Association (NFPA) is a great resource for fire statistics. Visit its website at *www.nfpa.org* to find current figures on numbers, causes, costs, etc.

The budgetary impact of wildland firefighting for the U.S. Forest Service came from a report dated August 2014: "The Rising Cost of Fire Operations: Effects on the Forest Service's Non-Fire Work."

Stephen Pyne is a guru of fire's role in human culture. *Fire in America* certainly served as a role model for this manuscript and I look forward to reading his next contribution based on his presentation at the 2013 Backyards and Beyond Conference in Salt Lake City, Utah.

2 – Fundamentals of Fire Science

The open quotations are from a pair of ancient stories. The first appears in A.R. Brown, *The Andaman Islanders: A Study in Social Anthropology* (London: Cambridge University Press, 1922), 202-203, which I found at *www.archive.org/details/ andamanislandersOOradc*. The second quotation and larger excerpt two paragraphs later are used with permission from Glenn Welker, owner of the Indigenous Peoples Literature Website *(www.indigenouspeople.net)*, and compiler of the stories posted there including "How Fire Came to the Six Nations."

I found the Cowichan story at *www. firstpeople.us/FP-Html-Legends/ WhoWasGivenTheFire-Cowichan. html*. For most stories found online, I was unable to find an author or website owner from whom to request permission to use the information. In those cases, I treated the story as a small part of a larger work and will credit the website in these notes.

The Yaqui story is from Harry Behn, ed., *Yaqui Myths and Legends Collected by Ruth Warner Giddings*, with illustrations by Laurie Cook (Tucson: The University of Arizona Press, 1959).

The Mewuk stories are from C. Hart Merriam, ed., *The Dawn of the World: Myths and Weird Tales Told by the Mewan [Miwok] Indians of California* (Cleveland: Arthur H. Clarke Co., 1910), and *www.firstpeople.us/FP-Html-Legends/How_Tol-le-loo_Stole_Fire-Miwok.html*.

I found the story of Maui in W. D. Westervelt, *Legends of Maui: A Demi God of Polynesia and of His Mother Hina* (Honolulu: The Hawaiian Gazette Co., Ltd., 1910).

The description of the flame is from Benjamin J. Ames, "What is a flame?" YouTube video published 4/26/12.

The excerpt from Colleen Morton Busch is from *Fire Monks*, 189.

Gregory Golgett, a professor at Eastern Kentucky University, presented fire investigation basics at a conference in Chelan, WA, in October 2014. I used my notes from that presentation to explain flashover, flameover and backdraft. If I missed a detail, the fault is mine rather than his.

I utilized the Internet to find accessible explanations of friction. One can be found at *www.discovery.com/tv-shows/mythbusters/about-this-show/friction-start-a-fire.htm*.

Primary sources can be difficult to find, but when found they are gifts. One of my favorites is Randolph B. Marcy, *The Prairie Traveler: The 1859 Handbook for Westbound Pioneers* (Mineola, NY: Dover

Publications, 2006); reprint of (New York: Harper and Brothers, 1859).

3 – Youth and Firesetting

The opening quotation about the Mewuk Chief is from Louis Barrett, *A Record of Forest and Field Fires in California from the Days of the Early Explorers to the Creation of the Forest Reserves* (U.S. Forest Service, 1935), 51, which was quoted in Stephen J. Pyne, *Fire in America: A Cultural History of Wildland and Rural Fire*, with a Foreword by William Cronon and a new Preface by the author, Cycle of Fire Series (Seattle: University of Washington Press, 1997), 71. The second quotation is from a King James Version of *The Bible*.

"Nookomis, it is very cold..." is from Gerald Vizenor, ed., *Summer in the Spring: Anishinaabe Lyric Poems and Stories, New Edition* (Norman, OK: University of Oklahoma Press, 1993), 104.

Quotations from participants in South Metro Fire Rescue's juvenile firesetter intervention program are used with permission from Program Director Cheryl Poage. All names have been changed to protect the identity of the child as well as the integrity of the file and the program itself.

Dr. Brad Jackson of Children's Hospital Colorado deserves credit for helping me understand the research of Jean Piaget and Lev Vygotsky and how it applies to youth fire misuse. Their theories are found in many instructional methodology and education psychology texts. Life safety educators can improve their craft by considering these theories.

Michelle Johnson generously shared the story of her daughter's experience with fireworks with me. That story has forced me to re-think how we teach fire safety. As a society we approach other potentially dangerous activities with education on how to act safely. It works with driver's education, hunter safety and even sex ed. Why don't we empower students and adults to use fire safely?

The American Pyrotechnics Association press release arrived in my email inbox somewhat serendip-itously, but the NFPA report by John Hall (2011) on fireworks safety and fires debunked what had been an awesome claim that relaxed laws on fireworks consumption and industry education were reducing injuries.

4 – Fire's Dark Side

See the King James Version of *The Bible* for the opening quotation from the Book of Joel. For the second quo-tation see Louis Hennepin and Victor Hugo Paltsits, *A New Discovery of a Vast Country in America*, Vol. I, edited by Reuben Gold Thwaites, (Chicago: A.C. McClurg & Co., 1903), 146, which is a reprint of the 1698 edition printed in London, England.

Again, the NFPA is a great resource for fire statistics.

Robert Gray has posted his research on several websites and presented his findings at conferences such as the 2013 Backyards & Beyond, Utah.

Details and quotations from the Sampler Mill Fire are from interviews that I conducted with participants while writing '*To Provide the Means of Protection': Volunteer Firefighting in Idaho Springs, Colorado, 1878-2003* (Idaho Springs, CO: Lulu Publishing, 2010), and from the *Clear Creek Courant* and *Rocky Mountain News*.

The letters from King and Hutchinson are from Paul M. Angle, ed., *The Great Chicago Fire* (Chicago: The Chicago Historical Society, 1946), 38, 17.

The San story appeared at *www. bigsiteofamazingfacts.com*, while a brief summary of the San culture can be found at *www.uiowa. edu/~africart/toc/people/San.html*.

The Maidu story is from Stith Thompson, *Tales of the North American Indians* (Cambridge, MA: Harvard University Press, 1929).

The Tolowa story is used with permis-sion from Katharine Berry Judson, ed. *Myths and Legends of California and the Old Southwest*, with an Introduction by Peter Iverson, Bison Books Ed. (Lincoln, NE: University of Nebraska Press, 1994), 68-69; reprint of Chicago: A.C. McClurg & Co., 1912.

I found the Nimipu story in Ella E. Clark, *Indian Legends from the Northern Rockies* (Norman: University of Oklahoma Press, 1966).

For the Aka-Jeru story, see Brown, *The Andaman Islanders*.

Warren Yahr generously gave me permission to use a story from his engaging book *Smokechaser* (Boise: Caxton Press, 1995).

Deanna Harrington, Deputy Fire Marshal for Arvada Fire Protection District (CO), shared this story with me in April 2015.

The quotations from the Peshtigo Fire are from Denise Gess and William Lutz, *Fire Storm at Peshtigo: A Town, Its People, and the Deadliest Fire in American History* (New York: Henry Holt And Company, 2002).

For the full account of how foresters viewed wildfire as a threat to re-sources, see *Report of the Committee Appointed by the National Academy of Sciences upon the Inauguration of a Forest Policy for the Forested Lands of the United States to the Secretary of the Interior* (Washington: Government Printing Office, 1897).

Clarence B. Swim's account of the Big Blow-up is in a report found at *www.fs.usda.gov/Internet/FSE_ DOCUMENTS/stelprdb5123029.pdf*.

Powell's quotation is from his article "The Non-Irrigable Lands of the Arid Region," *The Century Magazine*, XXXIX (November 1889–April 1890): 919.

My copy of *The Art of War* was published in 2003 by Barnes & Noble Classics. It has an introduction by Dallas Galvin and comes from the translation by Lionel Giles.

This phrasing of Samson's revenge is from the Book of Judges 15:4-5, *The Bible, King James Version*.

Greek fire remains enigmatic. I found older resources about its composition such as *A Concise History of Ancient Institutions, Inventions, and Discoveries in Science and Mechanic Art*, which was published in 1823 in London.

The quotation from Geoffrey de Vinsauf is from John de Joinville, *Chronicles of the Crusades: Contemporary Narratives of the Crusade of Richard Coeur de Leon* (London: George Bell and Sons, 1888), 115.

One version of the Harfleur Siege is found in John Hewitt, *Ancient Armour and Weapons in Europe from the Iron Period of the Northern Nations to the End of the Seventeenth Century* (London: John Henry and James Parker, 1860), 489.

The brief summary of napalm's history is from Mark Thompson, "Napalm: A True American Tale," at TIME.com. A thorough version of its history can be found in Napalm: *An American Biography?*, which was written by Historian Robert M. Neer.

"Create havoc..." is from an article James Powells wrote for *World War II* in 2003. It was quoted in Linton Weeks, "Beware Of Japanese Balloon Bombs," National Public Radio. None of the balloons ignited forest fires of the size desired by Japanese military leaders, but one did land in the woods near Klamath Falls, Oregon, and kill a Sunday School teacher and several kids.

5 – Fire's Positive Side

The opening quotations are from Robert J. Burdette, *The Drums of the 47th*, with an Introduction by John E. Hallwas (Urbana, IL: University of Illinois Press, 2000), and Carl Coke Rister, *Comanche Bondage: Dr. John Charles Beales's settlement of La Villa de Dolores on Las Moras Creek in Southern Texas of the 1830's with an annotated reprint of Sarah Ann Horn's Narrative of her captivity among the Comanches her ransom by traders in New Mexico and return via the Santa Fe Trail*, with an Introduction by Donald Worcester, Bison Books Edition (Lincoln: University of Nebraska Press, 1983).

This chapter begins with four ancient stories from Australia. They are from C. W. Peck, *Australian Legends and Tales Handed Down from the Remotest Times by the Autocthonous Inhabitants Of Our Land* (Sydney: Stafford, 1925); found at Project Gutenberg Australia eBooks *http://gutenberg.net.au/ebooks06/0607141.txt*; also *http://members.optusnet.com.au/virgothomas/space/abobeliefs2.html#TheFirst Fire*; and Roland B. Dixon, *Oceanic Mythology. Mythology of All Races Series*, Louis Herbert Gray, editor, Volume IX (Boston: Marshall Jones Company, 1916); Dixon, *Oceanic Mythology*.

The Karok story is from *www.firstpeople.us*.

The *Oceanic Mythology* book by Dixon also contains stories from the Motus and Dyaks.

That fire use among native cultures was as diverse as the cultures themselves was argued by Pyne in *Fire in America* and summarized in Julie Courtwright, "'When We First Come Here It All Looked Like Prairie Land Almost': Prairie Fire and Plains Settlement," *Western Historical Quarterly* 38 (Summer 2007): 161. Courtwright's article expertly combined multiple first-person accounts with current research into wildfire's ecological roles.

This quotation from John Wesley Powell appears in "The Non-Irrigable Lands of the Arid Region," 919.

Courtwright argues that bison herds and wildfire maintained the Great Plains as treeless habitat. Settlers initiated massive changes to the Plains when they obliterated the herds and excluded fire. See Courtwright, 161, and Kenneth F. Higgins, *Interpretation and Compendium of Historical Fire Accounts in the Northern Great Plains*, Resource Publication 161 (Washington, DC: United States Department of the Interior, Fish and Wildlife Service), 8.

"There used to be big fires..." is in Courtwright, 162-63. She pulled the quotation from Dick Rice, interview by Frank Benede, 2 June 1973, Oklahoma Oral History Collection, Oklahoma State Historical Society.

"Another says that..." is from Charles E. Bessey, "Are the Trees Advancing or Retreating Upon the Nebraska Plains," *Science* 10 (24 November 1899): 769. Bessey included several comments from farmers about trees newly growing in the area. Although this quotation was the only one that linked tree spread to fewer prairie fires directly, he referenced an earlier

article and many other interviews that expressly linked the two conditions.

"...cause springs to break forth..." and "or shall we be savages..." are in Courtwright, 171. She found the quotation in the *Walnut Valley Times* (El Dorado, Kansas) 20 October 1871, which had reprinted an article or editorial from the *Fort Scott Monitor* (Fort Scott, Kansas).

Statistics regarding prescribed burning is from "2012 National Prescribed Fire Use Survey Report" Technical Report 01-12 researched and published by the National Association of State Foresters and Coalition of Prescribed Fire Councils. It can be found at *www.prescribedfire.net*.

The Choctaw story is used with permission from Glenn Welker, owner of the Indigenous Peoples Literature Website (*www.indigenouspeople.net*), and compiler of the stories posted there. He received the story from Three Feathers, a resident of New Brunswick.

6 – Rules of Fire, Rites of Fire

This chapter's opening quotations are from Frank Russell, *Myths of the Jicarilla Apache*, Electronic Text Center, University of Virginia Library, *http://etext.lib.virginia.edu*; and "Instructions for Fire Prevention Inspectors of Bureau of Fire," Philadelphia Fire Department, December 1913.

More information regarding the transition from pedestrian to equestrian cultures can be found in Preston Holder, *The Hoe and the Horse on the Plains: A Study of Cultural Development among North American Indians* (Lincoln: University of Nebraska Press, 1974).

A brief summary of Neanderthal use of fire is found in the article

"Neanderthals Were Nifty at Controlling Fire," *ScienceDaily*, (15 March 2011).

The story of Tamboeja is from Dixon, *Oceanic Mythology*.

The Ife story is from John Wyndham, *Myths of Ífè* (London: Library of Alexandria, 1921).

The Lnu'k story is from W. H. Mechling, *Malecite Tales*, 1914, but I found it at *www.native-languages. org/maliseetstory.htm*.

The stories from the Alabamas are two of my favorites because they are so compact and instructive. They are both from John R. Swanton, *Myths and Tales of the Southeastern Indians*, Smithsonian Institution Bureau of American Ethnology Bulletin 88 (Washington, DC: United States Government Printing Office, 1929).

I found a brief biography of Marcy on 12 August 2013 at *www.kancoll. org/books/marcy/marcyaut.htm*; the excerpt is again from Marcy, *The Prairie Traveler*.

The complete diary of Warner is from Venola Lewis Bivans, editor, "The Diary of Luna E. Warner, A Kansas Teenager of the Early 1870s," 35 *Kansas Historical Quarterly* (Winter 1969): 294-5.

The story of Ti'iti'i is from Westervelt, *Legends of Maui*.

The history of the 10 and 18 is from A. W. Greeley, "Report of the Task Force to Recommend Action to Reduce the Chances of Men Being Killed by Burning While Fighting Fire," (US Forest Service, 17 June 1957).

I included the full presentation by Gleason because his insights continue to guide safety among wildland and other firefighters: Paul Gleason, "LCES," Presentation to the Fire and Aviation Staff of the US Forest Service, June 1991.

7 – Sacred Fire

The introductory quotations for this chapter are from Exodus 3:2 (King James Version) and an email from Rabbi Rob Cabelli, Associate Chaplain & Rabbi, Center for Religion, Spirituality & Social Justice, Grinnell College, Grinnell, Iowa, dated 23 December 2013.

This data on candle fires is from "Candle Fires in Residential Structures," Topical Fire Research Studies, Volume 6 (July 2006), Emmitsburg, MD: US Department of Homeland Security, US Fire Administration, National Fire Data Center. The US Fire Administration is comparable to the NFPA for fire data.

"I am the Light of the world..." appears in John 8: 12.

See the following three sources for additional information on the symbolism of candles: Friedrich Rest, *Our Christian Symbols*, with Illustrations by Harold D. Milton (New York: The Pilgrim Press, 1961); Marian Therese Horvat, "Votive Candles, Fire and the Love of God," *www.traditioninaction.org/ religious/f004rp.htm*; Rev. William Saunders, "The History of Votive Candles" *Arlington Catholic Herald*, no date.

"And thou shalt make..." appears in Exodus 25:31-40.

"Not by might..." Zechariah 4:1-6

"A fire devoureth..." Joel 2:3

"Then take of them..." Ezekiel 5:4

"And again I say unto..." is in the Book of Alma 5:52 within the *Book of Mormon*.

"Now Moses..." Exodus 3:1-6

"And has there come..." appears in TaHa 20:9-13 with the *Quran*. The story also appears in Al-Qasas 28, verses 29 and 30, although that source is not quoted here. Both were found at *www.noblequran.com*.

Gayanashagowa is available online at *www.mohawktribe. com/constitution/iroquois_ constitution_001.htm*.

"The lower part of his face..." is excerpted from Albert J. Carnoy, "Iranian Mythology," in *The Mythology of All Races* Vol. VI (Boston, 1917); found at *http:// rbedrosian.com/carn2.htm*. All references to Iranian traditions are from this source unless mentioned otherwise.

The five types of fire appear in Chapter 17 Verse 11 of the *Avesta Yasna*. I found a digital edition created by Joseph H. Peterson entitled *Avesta Yasna: Sacred Liturgy and Gathas/ Hymns of Zarathushtra* (1995), at *www.avesta.org/yasna/yasna.htm*.

"Thus, O Ahura..." appears in Avesta Yasna 34:4. Another good resource on Iranian traditions is No author, "Fire," Persian Culture Online, *http:// shahrukhfan7888.tripod.com/ persiancultureonline/id20.html*.

"He said..." appears in Al-Anbiya' 21:67-70 of the *Quran*, found at *www.alislam.org*.

When I began writing this chapter I sought subject matter experts. I figured finding Christian and Jewish contacts would be simple, but none of the pastors, priests or rabbis whom I contacted were willing to assist beyond a generic email. Thus I turned to books and the Internet for my research. My experience with an expert in Islamic traditions was completely different.

I reached out through Facebook to see if anyone I knew had a contact. My friend Beth (MacConnell) McNaughton had a friend in the Washington, DC, area whose father (Qusair Mohamedbhai) attended services at a mosque in the Denver area. I called Mr. Mohamedbhai and he gave me the contact information for the Imam at the Colorado Muslim Society's Denver Mosque. I called him, set an appointment for December 2013 and attended part of a service after being welcomed by the equivalent of ushers. When the service ended, I met with Assistant Imam Fudail Hassan. I remain grateful for his willingness to share his traditions with me. Any misinterpretations of his comments and his Islamic beliefs are mine alone although I did my best to bring his insight into this manuscript.

"Fire Is Part of Hell..." appears in Al-Baqarah 2:24 (*Quran*), *www. noblequran.com*.

"Thus has the word..." is from Al-Mu'min 40:7-8 (*Quran*), *www. alislam.org*.

"And according to the power..." is in Jacob 6:5 (*Book of Mormon*).

"And the righteous..." is in Section 29 Verses 27 and 28 of *The Doctrine and Covenants of The Church of Jesus Christ of Latter-day Saints, Containing Revelations Given to Joseph Smith, the Prophet with Some Additions by His Successors in the Presidency of the Church*.

I chatted with a coworker about the Jehovah's Witness interpretation of Hell in May 2015.

Learn about other forms of care and respect for fire within the Armenian tradition in Mardiros H. Ananikian, *Armenian Mythology*, 54, found at *http://rbedrosian.com/ananik4b.htm.*

The Alabama-Coushatta story is from Swanton, *Myths and Tales of the Southeastern Indians.* The excerpt containing their rules for fire-building is from R. E. Moore, "Texas Indians," *www.texasindians.com.*

The Creek story of rabbit is from Swanton, *Myths and Tales of the Southeastern Indians.*

The Zia story is from Katharine Berry Judson, editor, *Myths and Legends of California and the Old Southwest,* with an Introduction by Peter Iverson, Bison Books Ed. (Lincoln, NE: University of Nebraska Press, 1994), 68-69; reprint of Chicago: A.C. McClurg & Co., 1912; excerpt used with permission.

8 – Risk Perception and Fire

Both opening quotations are from Colleen Morton Bush, *Fire Monks.*

All quotations from residents of The Retreat subdivision are from a survey South Metro Fire Rescue Authority collected in September 2014. Answers were submitted anonymously. I didn't change any spelling, punctuation or grammar in the responses.

"Consider a turkey..." is from Nassim Nicholas Taleb, *The Black Swan: The Impact of the Highly Improbable* (New York: Random House, 2010), 40.

Detailed information on the Wallow Fire appears in Paul Keller, editor, "How Fuel Treatments Saved Homes from the 2011 Wallow Fire," Wildfire Lessons Learned Center, No date.

I have attended several presentations and read several articles penned by Sarah McCaffrey of the U.S. Forest Service. I suggest starting with her article "Thinking of Wildfire as a Natural Hazard," *Society and Natural Resources* 17:509-516.

9 – Harmony with Fire

The introductory quotations are found in Richard White, *Land Use, Environment, and Social Change: The Shaping of Island County, Washington,* with a Foreword by William Cronon (Seattle: University of Washington Press, 1980; James Malin, *History and Ecology: Studies of the Grassland,* Robert P. Swierenga, editor, (Lincoln: University of Nebraska Press, 1984).

"And least their firing..." is from Thomas Morton, *New English Canaan: Tracts and Other Papers, relating principally to the Origin, Settlement, and Progress of the Colonies in North America, from the Discovery of the Country to the Year 1776,* Peter Force, Collector, Volume II (Washington, DC: Peter Force, 1838); Originally published as its own book in 1637.

"Their method of clearing..." appears in John Sergeant, "History of the Muhheakunnuk Indians," an article within Jedidiah Morse, *The First Annual Report of the American Society for Promoting the Civilization and General Improvement of the Indian Tribes in the United States* (New Haven, CT: S. Converse, 1824), 41.

Other examples of how our continent's first residents utilized fire are in Thomas R. Vale, editor, *Fire, Native Peoples, and The Natural Landscape* (Washington, DC: Island Press, 2002).

"Rather than being..." is in White, *Land Use,* 21.

The quotation from James Cooper is from *Reports of Exploration and Surveys to Ascertain the Most Practicable and Economical Route for a Railroad from the Mississippi River to the Pacific Ocean* Volume XII Book II (Washington, DC: Thomas H. Ford, 1860), 23.

"Prairie fires were not..." is in Laurie Arnold, "Recollections of Lizzie Anna Dopps as told to her daughter Jessie E. Botsford," Norton County Kansas GenWeb website, *http://skyways.lib.ks.us/genweb/norton/dopps_diary/preface.htm.*

"Workers killed a cow..." is from Courtwright, 167. She paraphrased the icky process from Esther Dellis, "Recollections of Joe Killough: An Interview With a Pioneer," (1941), in the Panhandle Plains Historical Museum Archives, Canyon, Texas, 5, and Roscoe Logue, *Under Texas and Border Skies* (Amarillo, TX: Unknown Publisher, 1935), 44.

The NFPA's new version of a WUI is found in Molly Mowery, "The WUI Revisited," *NFPA Journal* (1 July 2013): 1.

The reciprocal interactions quotation is from Mark Fiege, "Creating a Hybrid Landscape: Irrigated Agriculture in Idaho," *Illahee* 11 (Spring-Summer 1995), 60. Other references to Fiege also are from this article.

"That man is, in fact..." is from Aldo Leopold, *A Sand County Almanac and Sketches Here and There*, with illustrations by Charles W. Schwartz and an introduction by Robert Finch (New York: Oxford University Press, 1987), 205.

All of the references to Carney's work is from Elizabeth Carney, "Suburbanizing Nature and

Naturalizing Suburbanites: Outdoor-Living Culture and Landscapes of Growth," *Western Historical Quarterly* 38 (Winter 2007): 478, "where Rocky Mountain penstemons..." is from 499.

For Census data I turned to Frank Hobbs and Nicole Stoops, *Demographic Trends in the 20th Century*, Census 2000 Special Reports series (November 2002).

Karen Hajek emailed me her memories of the Waldo canyon fire in May 2015.

I pulled the quotation from a resident of Black Forest from an email sent by the Pikes Peak Wildfire Prevention Partners in July 2013.

All quotations from James Malin are from *History and Ecology*.

The quotation from Jack Ward Thomas was quoted in Rocky Barker, *Scorched Earth: How the Fires of Yellowstone Changed America*, (Washington, DC: Island Press, 2005).

The enormous waste..." is in *Report of the Committee Appointed by the National Academy of Sciences*, 8-9.

"A right circular cone..." is in Francis Weston Sears and Mark W. Zemansky, *University Physics: Complete Edition*, 2d ed., with Supplementary Problems (Reading, MA: Addison-Wesley Publishing Company, 1955), 19. This book is my source for Le Chatelier's Principle as well.

"The role of *Homo sapiens*..." is in Leopold, *A Sand County Almanac*, 204.

Physicist Neils Bohr is credited with the comment, "Prediction is very difficult, especially about the future."

10 – Will We Keep Burning?

The opening quotations are from *The Works of Francis Bacon, Lord Chancellor of England*, A New Edition with a Life of the Author by Basil Motagu, Vol. I (Philadelphia: Carey and Hart, 1842), 431; and Colorado Governor John Hickenlooper, to Ryan Warner on "Colorado Matters," Colorado Public Radio KCFR, 25 June 2013.

The story of Goorda is in "Folktales from Around the World," 2006 supplement to *The Denver Post*. Commissioned from Hot Topics Publications by Jill Scott, Newspaper in Education Manager. Stories reprinted with permission from Fulcrum Publishing. Reprinted from *Earth Tales from Around the World* by Michael J. Caduto, illustrated by Adelaide Murphy Tyrol (Golden, CO: Fulcrum Publishing, 1997).

The Dutch efforts toward fire prevention are described in Edward C. Goodman, *Fire!: The 100 Most Devastating Fires Through the Ages and the Heroes Who Fought Them* (New York: Black Dog & Leventhal Publishers, 2001), 19; Boston's efforts figure prominently in Peter Charles Hoffer, *Seven Fires: The Urban Infernos that Shaped America* (New York: Public Affairs, 2006), 23.

Visit *www.nfpa.org* for a detailed history of Fire Prevention Week.

The decisions, conversations and presentations of the 1913 conference are captured in Powell Evans, ed., *Official Record of the First American National Fire Prevention Convention* (Google Book search, 2009), which is a reprint of (Philadelphia: Merchant & Evans, 1914).

President Coolidge's remarks are from *www.publicsafety.net/*
prevention.htm while President Truman's comments were in a press release dated 3 January 3, 1947 (copy in possession of the author).

President Nixon's comments were in The National Commission on Fire Prevention and Control, *America Burning* (Washington, DC: National Commission on Fire Prevention and Control, 1973), i.

"I spoke last week..." is a longer excerpt from Hickenlooper, 25 June 2013.

The National Center for Injury Prevention and Control published *Global Concepts in Residential Fire Safety* in October 2007 and August 2008.

For the full list of Life Safety Initiatives, visit *www. lifesafetyinitiatives.com.*

Lisa Miklas shared part of her experience with me via email on June 3, 2015.

A history of smoke detection is in "Home Smoke Alarms and Other Fire Detection and Alarm Equipment," White Paper Series, Public/Private Fire Safety Council (April 2006): i; and *www.whoinventedit.net*. Another resource is a Fact Sheet produced by the U.S. Nuclear Regulatory Commission.

The residential sprinkler reports referenced in the text are Jim Ford, *Saving Lives, Saving Money: Automatic Sprinklers, A 10-Year Study* (Scottsdale, AZ: Rural/ Metro Fire Department, 1997); and "Communities with Home Fire Sprinklers: The Experience in Bucks County, Pennsylvania," (Washington Crossing, PA: Home Fire Sprinkler Coalition, 2011).

Glossary

arson. The act of deliberately setting a fire.

backdraft. The explosive fireball that results from incomplete combustion within a building. When a limited supply of oxygen causes combustion to produce large volumes of unburned vaporized fuel, the introduction of oxygen to that vapor creates a volatile mixture of gases. If contacted by a source of sufficient heat, extremely rapid burning can result, forcing a fireball out the opening through which oxygen has just entered.

BTU or Btu. A British thermal unit is the amount of energy needed to heat one pound of water by one degree Fahrenheit.

carbon dioxide. A toxic byproduct of fire capable of causing illness or death.

carbon monoxide. An odorless and colorless deadly byproduct of fire that displaces oxygen in our bloodstream.

combustion. The rapid chemical combination of a substance with oxygen, involving the production of heat and light.

conduction. The movement of energy through a substance from particle to particle in response to a difference in temperature. As energy moves from particle to particle, the particles do not move.

convection. The movement of a gas or liquid in response to a temperature difference.

ember. An airborne piece of burning material that can spread a fire beyond its original area. Most embers are pea-sized or smaller, but larger fires can launch larger embers into the air. They are usually responsible for burning homes in wildfires.

flameover. When flames roll through superheated particles of smoke in billowing waves.

flashover. When all surfaces exposed to thermal radiation reach their ignition temperatures nearly simultaneously and catch fire. Nothing survives flashover.

fire. The popular term for combustion or burning, in which substances combine chemically with oxygen from the air and typically produce light, heat and smoke.

fire interest. Marks the first attempt of children to learn about fire, usually taking the shape of questions and play, such as wearing plastic fire helmets and pretending to cook with toy stoves. Kids age 3–5 are learning about their environment and experiment with how to control those surroundings.

firesetting. When children seek ignition sources, fuels and accelerants and use fire to express themselves and their needs. If fires relieve anxiety or fear, the children will set more.

firestarting. When children experiment with ignition sources such as matches, lighters and magnifying glasses. With supervision, these children learn appropriate and safe ways to interact with fire. Without supervision, they can hurt themselves and start unintentional fires.

fire triangle. Fire needs three ingredients in order to burn: heat, fuel and oxygen. A triangle represents their relationship because when any one ingredient is removed, the fire stops.

flames. Glowing soot particles. Efficient burning produces flames of blue light; inefficient burning produces flames of yellow, orange and red.

friction. The resistance that one surface or object encounters when moving over another, which produces heat.

fuel. Although it is used most often to describe petroleum projects, a fuel is any material capable of combustion.

greek fire. A weaponized form of fire used throughout the Mediterranean region between the 7th and 13th centuries.

handcrew. A group of wildland firefighters who train together and work together to extinguish wildfires.

heat. In physics, a form of energy associated with the movement of atoms and molecules in any material.

hózhǫ. A Navajo term for living in harmony with one's surroundings. I use it to represent a new vision for human relationships with fire.

mitigation. Actions that reduce the severity, seriousness or painfulness of something. Wildfire mitigation consists of modifying vegetation and making homes more fire-resistant.

napalm. Naphthenic palmitic acid is a weaponized form of fire that uses a burning gel to destroy buildings and organisms.

PPE. Personal Protective Equipment protects firefighters from heat and smoke. PPE also describes hard hats, non-latex gloves, harnesses and other life-protecting equipment.

prescribed fire. Often used by land managers, firefighters and other scientists to modify ecosystems to prevent larger wildfires and to change vegetation growing in certain areas.

radiation. When waves of infrared electromagnetic radiation travel from a liquid or solid through the air without heating the air itself. As the waves are absorbed by another liquid or solid object in line of sight with the heat source, energy is converted to heat.

reactant. A substance that takes part in and undergoes change during a reaction, such as combustion.

residential sprinkler system. Life-saving technology that sprays water to contain fires and provide more time for humans to escape burning buildings. Individual sprinkler heads only react to heat from a fire.

SCBA. Self-contained breathing apparatus allows firefighters to breathe clean, cool air in buildings filled with toxic or superheated air.

smoke. Countless tiny particles of partially burned fuels released during combustion, most of which are hazardous to humans. Smoke also tends to be hot, which can hurt humans and spread fire in advance of flames.

soot. Carbon atoms that bond together after being released during combustion.

temperature. The measurement of heat in a given substance.

thermodynamics. The study of how heat moves.

wildland fire or wildfire. Any fire that burns vegetation.

WUI (Wildland Urban Interface). An area that combines human structures with native vegetation. These fires receive strong attention from firefighters and the public because they threaten lives and property.

youth fire misuse. Previously known as juvenile firesetting. Occurs when children use fire inappropriately. Each state sets its own rules regarding consequences for such fires. In Colorado, children age 10 and older who misuse fire commit arson.

Index

Author Einar Jensen

$\mathcal{E}$inar Jensen, a veteran community risk reduction specialist, works for Colorado's South Metro Fire Rescue Authority's Preparedness Division directing their wildfire mitigation program and educating students and the public on fire and injury prevention. Prior to leaving the mountains, he was a volunteer firefighter for Clear Creek Fire Authority for twelve years and, concurrently, volunteered for eight years for Clear Creek Sheriff's Office Marmot Wildfire Crew. He studied environmental history at the University of Montana and University of Puget Sound and considers himself a recovering print journalist. He is an active member of the Fire & Life Safety Educators of Colorado.

100% solar & wind powered since 1999

PixyJack Press INC.

Visit **PixyJackPress.com** to view all of our titles, including *Surviving Wildfire*, and to place wholesale orders for your school or organization.

www.ingramcontent.com/pod-product-compliance
Lightning Source LLC
Chambersburg PA
CBHW021924190326
41519CB00009B/901